When it Rains, Does Space Get Wet?

When it Rains, Does Space Get Wet?

Living the Time-Space-Knowledge-Vision

Jack Petranker

Dharma Publishing

PERSPECTIVES IN TIME SPACE AND KNOWLEDGE

Dimensions of Thought I and II
Mastery of Mind
Visions of Knowledge
Light of Knowledge
A New Way of Being
A New Kind of Knowledge
When it Rains, Does Space Get Wet?

NYINGMA EDUCATION SERIES

Ways of Enlightenment
When It Rains, Does Space Get Wet?

Library of Congress Control Number 2006900306

ISBN 0-89800-387-3

Cover photograph courtesy of the National Oceanic and Atmospheric Administration Central Library Photo Collection.

Typeset in Trump Mediaeval, Helvetica Light, and Bauhaus.
Printed and bound by Dharma Press, Berkeley and Cazadero, CA.

9 8 7 6 5 4 3 2 1

Dedicated to the

flowering of creative inquiry

Contents

TSK Guide: 75 Topics

Nine Levels in Six Books

Three Levels In Sixteen Aspects

Summaries of the Books

Appendices

Preface

This book is a companion to the Time-Space-Knowledge vision, as presented in six books written by Tarthang Tulku between 1977 and 1997. It is meant both for students of the TSK vision, and for those who would like a way into the vision. Since it is so closely linked to works in the TSK series, it comes with a CD (ROM) that includes the complete, searchable text of all six books, formatted to match the original texts. The CD includes the software necessary to view and search the books, and should be very useful for serious students.

When it Rains is intended as an invitation to inquiry, a support for anyone interested in the TSK vision. The vision has much to offer, both for individuals and for our increasingly dysfunctional society. My hope is that this book will help make its riches more widely available and deepen the appreciation of those who already know the vision's value.

Overview of the Book

When it Rains is organized into six sections, followed by several appendices. Here is a brief overview:

> *Study Program in the TSK Vision* is a 36-unit program, designed as an integrated, sequential whole. It covers the TSK books in some detail, discussing major themes and giving readings and exercises for each unit. An online

class based on the Study Program is planned; for information, see the Dharma Publishing website.

TSK Guide: 75 Topics This section organizes the themes of the TSK vision according to the three levels of time, space, and knowledge, making nine divisions in all. For each topic, references are given to the central place where that topic is introduced or discussed, together with related readings and exercises. The aim is to create a resource that readers and students of the TSK vision can approach in their own way and at their own pace, letting their interests and their sense of what seems appropriate guide them. The topics might also be useful in organizing TSK workshops or classes. For a list of the 75 topics, see p. 158.

Nine Levels in Six Books This section presents the three levels of space, time, and knowledge (again, nine levels in all), using them as an organizing principle to analyze in a preliminary way the contents of all six TSK books. Much more could be done along these lines.

Three Levels in Sixteen Aspects analyzes the TSK books in the form of sixteen charts, one for each chapter of the TSK book. The tables single out the three levels of time, space, and knowledge as they relate to that chapter and give quotations meant to epitomize the approach of each book.

Summaries of the Books With so much material available on the TSK vision, these summaries may help orient readers and let them identify key themes as they emerge and unfold from one volume to the next.

Appendices There are four appendices: a collection of references to TSK methodology; a discussion of the characteristics of each of the TSK books (including their likely audience, their special strengths, and obstacles readers may encounter in dealing with them); an alternate study program that I developed more than ten years ago; and a list of articles by Tarthang Tulku on the TSK vision published in unexpected places.

Acknowledgments

First and foremost, I express my debt to Tarthang Tulku. Without him this book would not exist, and without the countless hours he has spent with me presenting and clarifying the TSK vision, I could never have imagined writing it. In particular, I thank him for suggesting several years ago that I prepare this work, and for inviting me to submit it to Dharma Publishing for publication.

My thanks go as well to Arnaud Maitland and Caroline van Tuyll, directors of Dharma Publishing, for their enthusiastic response to the idea of publishing this book, even before they had any inkling what it would contain.

Steve Randall, whose devotion to the TSK vision over the years is unmatched, put in the many hours necessary to create the TSK CD that forms part of this project. I am sure it will be a valuable resource for years to come. Zara Wallace long ago prepared summaries of the TSK books that I consulted in preparing the summaries found in this volume.

Finally, my thanks to the students with whom I have shared the TSK vision over the years, starting with my first halting efforts in Germany in 1983; to the teachers with whom I studied TSK in the early days after the vision had appeared, and to my co-teachers in the years since; to the editors who have joined with me in working on the books of the TSK vision; to the authors who contributed to the Perspectives on TSK volumes that I had the good fortune to compile; and to Leslie. I have learned much from them all.

Abbreviations

TSK	Time, Space, and Knowledge: A New Vision of Reality
LOK	Love of Knowledge
KTS	Knowledge of Time and Space
VOK	Visions of Knowledge: Liberation of the Modern Mind
DTS	Dynamics of Time and Space: Transcending Limits on Knowledge
SDTS	Sacred Dimensions of Time and Space

Study Program

in the

TSK Vision

Orientation

The aim of this program is to encourage those who work with it to accept the invitation that the Time-Space-Knowledge vision offers: "We can develop an intimacy with time, space, and knowledge, with consequences that will show up clearly in our lives." DTS xxi (For abbreviations, see p. x.)

You might think it would be easy to accept this invitation, because we cannot help being intimate with time, space, and knowledge. They are the stuff of our lives. Yet we have learned to turn away from this intimacy, and recovering it requires effort. In turn, this effort requires the wish to change something fundamental about our lives and way of being. If that wish is present in a sustained way, we will act on it, and then change is not that difficult.

This Study Program is meant to provide a comprehensive introduction to the TSK vision. You may just want to read around in it and see what you can pick up that seems useful or interesting, but my intention has been different. I have aimed to offer a systematic course of study, one that will require a certain discipline to follow. You can see for yourself if this approach speaks to you.

If you complete the whole program and do all the recommended reading, you will have read about 80% of TSK, major portions of LOK, Part One and other selected chapters from

KTS, Part One and other selected chapters from DTS, and smaller excerpts from VOK and SDTS. In all, you will have read about 40% of what Tarthang Tulku has written on TSK, and you should get a good sense of the special qualities of each of the texts.

Most of the books that present the TSK vision make use of specific exercises, and working with these exercises—or other exercises that can be found in the text or that you make up yourself—is a vital part of engaging the TSK vision. Try to do a TSK practice every day, for at least 20 minutes or so, including time for journal writing afterwards. If you can do more, that is even better: perhaps an hour or more, either in one sitting or in two to three sessions. Of course, you can also be doing practice as you go about your life. If you get involved in the program, you will probably find yourself doing this naturally.

How the Program Works

The program consists of thirty-six units. Each unit is based on suggested readings and exercises. As with the TSK vision itself, the exercises are integral to the program. With one or two exceptions, they are drawn from the TSK books. Almost every unit focuses on two exercises, and most units also suggest supplementary exercises for those who want to expand their inquiry and have the time to do additional practice.

The readings range from about 25 to 50 pages a week. The program is fairly structured, as you will see from this introduction. Even though this much structure may seem to go against the open and inviting quality of the TSK vision, I recommend that you try to stay within the structure. Freedom makes the most sense when it unfolds within limits.

If you decide to do the whole program and cover one unit a week, the program will take eight months to complete. Of course, you are free to proceed at your own pace. Despite what I said above, I do not recommend skipping around, since the program has its own logic and structure. If you do at least the first nine units, you should have a good sense of whether you want to continue with this approach.

The program is divided into three parts. Parts One and Two each contain nine units, while Part Three contains eighteen units. Part One presents the 'Body' of the vision, Part Two presents the 'Experience' of the vision, and Part Three presents the 'Knowledge' of the vision. These divisions, however, are only meant to be suggestive.

To give you an idea of the structure for the program, here is a brief survey of Part One. Units 1–3 are an introduction, based on readings drawn mainly from the front matter for the six TSK books. The aim is to introduce basic themes and orientations characteristic of the TSK vision, without going into the specifics of the vision itself. This part of the program introduces a set of "fundamental exercises" that you can work with throughout the program (more on this below).

Units 4–6 begin to explore the vision in terms of the three fundamental facets of time, space, and knowledge. The themes are embodiment, threefold linear time (past-present-future), and opening perspectives.

Units 7–9 continue with the initial presentation of the vision. This time the themes are the space of thoughts, alternatives to linear time, and challenging models and interpretations. Repeating the basic space-time-knowledge structure in this way is a reminder that the themes of TSK operate on many levels simultaneously, and that it is possible to explore one theme from multiple perspectives.

Integrating Readings and Exercises

As in the TSK vision itself, this study program draws extensively on both readings and exercises, two closely integrated aspects of the vision. As you will see, some of the exercises assigned in the program are in fact not formally identified as exercises at all; instead, they are based on a passage or an image from the readings.

More generally, reading TSK with care and appreciation is itself an exercise. That is, we do not read TSK for the conceptual insights it offers, though these are important, but because as we read, something shifts at the level of experience. New possibilities open, new ways of engaging our lives manifest, and we find ourselves questioning what is happening in different ways. There are people who work with the TSK vision mostly through reading the books over and over again, not paying too much attention to the exercises. There are also people who come at the vision from the other side, practicing the exercises intensively without relying on the texts. In this program, we will aim to integrate the two.

Whether in the reading or in the exercises, you are bound to come to places where you do not understand what is being said or what you are supposed to do. It is important to keep in mind (as the TSK books say repeatedly) that it is not necessary to understand everything. Sometimes understanding comes as you engage the material and let it sink in. Sometimes you may never 'understand' in the way we usually imagine, but something fundamental will shift anyway.

Similarly, you do not need to feel obligated to do the exercises in "the right way." Sometimes you will not be able to follow the exercises just as they are written, but you will find that they have a powerful impact nonetheless. Sometimes an exercise will change as you practice, until you find yourself conducting an experiential inquiry that is only loosely connected with the instructions given in the text. And certainly you will find with every exercise (and for that matter, every reading) that it is different each time you do it.

As a rule, each unit of this program introduces two main exercises and one or two supplementary exercises. The supplementary exercises are for those of you who have the time to go into the program more deeply. However, if you find that you have the sense of being stuck with one of the main exercises, feel free to substitute a supplementary exercise. For the most part, I will not comment on supplementary exercises, but of course many of them have their own commentary in the texts.

Foundation Exercises

The six exercises assigned in the first three units of the Study Program are "Foundation Exercises," chosen because they present fundamental elements of the TSK vision in a way that allows for deepening over time. You should feel free to continue to explore these exercises throughout the program, in addition to the other exercises assigned for each unit.

The foundation exercises also serve a very practical purpose: If you find it difficult to engage a particular exercise assigned later in the program, you can use one of the foundation exercises as a kind of warm-up, a way to involve yourself in the practice. They can serve this function in part because each of them invites you to work with your experience just as it is, not trying to change it any way.

Although the program itself is quite ambitious, you can use the foundation exercises to bring to it another dimension of practice. Here is a suggested approach: At the end of the first three units, choose one of the foundation exercises that seems either particularly fruitful or particularly challenging and "specialize" in it by practicing it for two hours during the week in addition to the other practices assigned in the program. Continue with this 'specialized' practice through Part One of the program.

At the start of Part Two, switch to another foundation exercise, one that relates to a different facet of the vision (time, space, or knowledge). At the beginning of Part Three, choose a third foundation exercise, relating to the third facet of the vision, and do that exercise while you go through the next six units (units 19–24). For the following six units (25–30), return to the foundation exercise practiced during Part Two. During the final six units (31–36), go back to the foundation exercise done during Part One, coming full circle. Of course, it will help if you keep good journal notes on your practice of these exercises. The journal is discussed below.

Here for reference are the six Foundation Exercises:

Space

Ex. A	Expanding and Condensing
TSK Ex. 7	Body-Mind-Thought Interplay

Time

TSK Ex. 24	Marriage of Sound and Breath
LOK Ex. 5	Disowning Tension

Knowledge

DTS Ex. 1	Conducting the Vision
TSK Ex. 30	Subject-Object Reversal

Integrating Relaxation into Practice

Most students of the TSK vision find it useful to integrate relaxation exercises or simple stretching into their practice periods. TSK programs at the Nyingma Institute and its affiliated centers often use Tibetan Relaxation (Kum Nye) for this purpose. These exercises can be found in three books by Tarthang Tulku: *Kum Nye Relaxation*, *Tibetan Relaxation*, and *Joy of Being*, available through Dharma Publishing. Here is a list of exercises from *Kum Nye Relaxtion* that link well to the themes of space, time, and knowledge:

Space	1, 2, 3, 4, 5, 6, 24, 25, 26, 36, 42
Time	3, 14, 15, 17, 21, 24, 25, 33, 64, 27, 37, 42, 44, 46, 81, 97
Knowledge	22, 24,29, 44, 63, 69

This list is by no means complete; for instance, almost any of the balance exercises in Kum Nye can be used in relation to time. Again, many exercises can be related to all three facets of TSK. For instance, Kum Nye ex. 20 (Flying) can be done with a focus on the movement through space, on loosening the model of 'experience-as-linear-sequence', or on the tendency to take limiting positions.

Keeping a Journal

Keeping a journal is an important part of this program. Anyone who approaches the TSK exercises with an open mind will have rich and rewarding experiences, but if you simply file these experiences away mentally, you may miss the opportunity for sustained inquiry that the TSK vision offers. The vision will have the greatest power if you let it become a part of your life, actively pursuing TSK inquiry at all times. That is the point of a journal. Developing the habit of keeping entries in a journal can help you find ways to formulate your experience. It can also encourage you to look for connections, similarities, and so on. Finally, writing about your experience is itself an exercise. There is always a danger that writing will encourage a conceptual approach to experience, but this risk—the tendency to conceptualize—can itself become a theme for inquiry.

The journal can also help you formulate questions and go more deeply into places where you feel stuck. As with so many parts of the TSK vision, it is the times when you do not know how to proceed, the times when you think you might want to to quit or go off and do something else, that the vision may be about to open up. With this in mind, committing to the discipline of regularly making entries in a journal can be a very fruitful—and engaging—approach to taking your practice and inquiry to a deeper level.

Many people like to keep journals in a special bound book, but that may not work well for this program, because it presupposes a linear structure: this page following that page, this day leading on to the next. TSK inquiry is multidimensional, and this program trades on that multidimensionality.

We will be returning to the same themes, the same readings, and the same exercises at various times during the program, looking at them from different perspectives, holding them up to the light from different angles. To reflect this multiplicity, it probably makes the most sense to keep your journal in a loose-leaf binder, or else as a set of computer files.

You can organize the journal any way you like, and in fact you may want to change the organization as you go along. One approach would be to have three sections:

Readings These could be divided into the areas of space, time, and knowledge, and perhaps within that by book. This could include notes taken while reading, cross-references to experiences with specific exercises or to daily experience, quotations from the book, and questions that come up during reading that you then go on to explore in other ways.

Exercises Assign each exercise its own page or pages, and then take notes whenever you do an exercise. You may want to have a special section for the foundation exercises.

Daily Experience The aim of this Study Guide is to help you make the TSK vision a part of your life, so naturally you will want to reflect on how deepening awareness of, or interest in, the time-space-knowledge interplay is showing up in daily life. In part this means doing exercises while carrying out other activity, but in part it simply means being aware of the interplay, observing patterns, conditioning and positioning, noticing the self at work, and so on. Here we are talking about a journal in the truest and most literal sense: a daily record of what happens.

Practice Days and Retreats

If possible, set aside time at regular intervals for more intensive practice. Perhaps you could do this by organizing a study group, so that you have the support of studying with others. Given the structure of the program, it would be helpful to set intensive practice days after units 3, 9, 18, and 27, but I have suggested some other possible places along the way as well. A longer retreat at the end of the program will help you return to what you have been investigating and take it deeper. You can find some brief suggestions for a retreat in TSK, xviii.

If you do decide to set aside time for intensive practice, you will probably have some strong ideas of your own on how you would like to practice. In general, it is good to focus on relatively intensive practice of several exercises. These could be the foundation practices, or you might consider working with any of the following exercises:

First practice day: TSK Ex. 7, 8, 9, 10, 15, 16; KTS Ex. 36

Second practice day: any of the above, and TSK Ex. 14, 22, 23, 24, 25, 26, 30, 31, 34; DTS Ex. 6

For guidance on a very intense retreat experience focused on one exercise, take a look at TSK Ex. 15, A Mountain Retreat.

Unit One
The TSK Vision

VOK pp. vii–xix; TSK pp. xxxv–xli; SDTS p. 2;
KTS pp. xv–xxi; LOK pp. xxiv–xxxiv; DTS pp. xi–xxiv
Exercises: Ex. A (Expanding and Condensing);
TSK Ex. 24

This unit introduces the TSK vision. Because it sets the stage for what follows, I have more to say about the readings and exercises in this unit than I will in the units that make up the rest of the study program.

Usually an introduction leads you into something. (That is the literal meaning of the word "intro-duce.") But when it comes to the TSK vision (or perhaps any vision), that may be too linear a way of thinking about things. Instead, consider this introduction an opportunity to reorient your thinking and to begin looking at your ordinary experience in light of the possibility that it might all happen very differently. One way to describe this is to say that in this opening unit we are "warming up" to the Time-Space-Knowledge vision.

This sense of introducing yourself to the vision holds even if you have been studying the TSK vision for years and have read all the books several times over. We all tend to turn our insights and convictions into rigid structures, so it is always useful to start fresh, to exercise "beginner's mind." That is what the vision is about.

The readings for this unit come from the front matter of the various TSK books. Among the points they introduce are the following:

☆ The self is not the owner of knowledge or experience; nor is the self ultimately the one who experiences;
☆ Knowledge is an active force, inherent in what we know rather than imported from outside;
☆ TSK can be understood as a "methodology" for questioning and challenging conventional understanding;
☆ The nature of vision, and what it means to present a "new vision of reality;"
☆ Time, Space, and Knowledge as the being of our being.

All of these points will be dealt with repeatedly throughout the program, so there is no need to "master" them on this reading. They are listed here so you can check in on them, and perhaps see which ones speak to you most directly.

More important to activating the TSK vision than any particular insight is instilling a sense that we are embarked on a journey of discovery. Tarthang Tulku (I will follow Tibetan practice and refer to him as Rinpoche, a traditional title of respect) emphasizes this point again and again in the readings. He is at once generous and insistent:

> I do not own this vision. You are free to make your own commentaries, add your own insights, develop your own versions. For the sake of your growth and the well-being of your friends, I hope that you do so. I say this as more than a polite invitation. It is really up to you to activate this vision in your own life. If knowledge does not lead to more knowledge, what value does it have? If we cannot intensify our being alive, what is the point of our appearing in space? If our time is not productive and meaningful, it is time to move on. DTS xxiii

Inviting Inquiry

As Rinpoche writes in VOK, ix, the TSK vision relies on inquiry as a path to knowledge.This might not seem anything out of the ordinary, since our whole society, and especially science, are constantly engaged in inquiry. But the fact is that although we do not usually realize it, we are always imposing barriers to inquiry. The TSK vision challenges these barriers by challenging the models, the structures, and the positions that they take for granted.

Another way to say this is that in the TSK vision there are no limits on inquiry. Of course, there are things we do not and cannot know, at least in any ordinary way. 'Right now' as I write, I cannot know anything about your situation 'right now' as you read. But I can ask questions. For instance, I can ask what it is that separates us (me as writer and you as reader), and what connects us. Here are just a couple of ways to pursue that inquiry:

How is my 'right now' different from your 'right now', and what accounts for this difference? Perhaps the answer seems obvious: your present is my future, and my present is your past. But what does this mean? How does a past connect to a future? If you are located at one point in time and I am located in another, which is 'really' the present? Is one more real than the other?

Even though I know nothing about you in the sense that I do not know who you are, there are still some things I do know. I share with you the fact of being human, and this gives me some appreciation for who you are. If I didn't have that appreciation, there would be no way to write this paragraph. More specifically, you are a human being who knows how to read, and while I may not have thought much about it, certainly familiarity with the process of reading shapes your mind and makes it in some ways akin to mine. Finally, you have taken the trouble to read this far in the book, which gives us still more things in common.

In discussing the idea that inquiry has no limits, VOK states that knowledge is not what we think it is. Here is another point for inquiry. At this point in the text, VOK introduces the term "knowingness," a word that has no standard meaning in modern English. We will be talking about knowingness later in the program, but here is a little exercise: Use the search engine on the CD that comes with this book to look up "knowingness" and browse through some of the related discussions. One clear advantage of the term is its direct link to "knowing." Unlike the term "knowledge," knowing is an activity. Like inquiry, it remains unfinished. Take a look also at the discussion starting at the bottom of VOK xi, contrasting theoretical knowledge with the visionary knowledge of the TSK vision.

Although there are no limits on inquiry, we act as though there are. By acting in this way, we impose such limits. Why do we do this? "When we turn away from inquiry, what dynamic is at work?" (VOK ix) In other words, why do we not ask questions? What keeps this from happening?

Assume for a moment that we really have imposed unnecessary limits on knowledge. What are the consequences? VOK, p. xiv suggests an answer when it speaks of the *urgency* of our situation. Whether we recognize it or not, we are in trouble, living in limited ways and producing limited results. The more we can realize that something needs to change, the better, because then we will have the motivation to change. It is like someone who has been told she has cancer; she has a strong motivation to find out everything she can about available treatments, seek out other opinions, and experiment with alternatives to the conventional medical system.

It would be a good idea to reflect on your own motivation for working with this study program. Without a motive that keeps you going, sooner or later you will lose interest in asking questions. Is it the sheer pleasure in seeing things in new ways? Is it because you are looking for something in particular? These are questions only you can answer.

Asking questions is at the heart of the TSK vision, because visionary knowledge can never come from the presupposed:

> . . . their answers cannot be our answers. . . . We must each individually take up the challenge of knowledge for ourselves. VOK x

Not only that, but we cannot accept the answers and insights we may have discovered in the past, not even the ones we cherish and hold most dear. The insight that seemed so fruitful yesterday may feel lifeless today. Each time we ask, we have to ask anew. When we can take this ongoing evolution to heart, it can bring us deep joy.

Conducting Inquiry

Earlier it was suggested that TSK provides a "methodology" for inquiry. The readings qualify this statement in a useful way, making clear there is no one way to conduct TSK inquiry:

> No matter what form they take, questions evoke knowledge. We can foster this knowledge dynamic by expanding the inquiry, feeding questions into questions, asking with the heart, the senses, and the body as well as the mind. We can be free and open in our inquiry, for knowledge is not confined to one form, nor are space and time restricted in the possibilities they allow and activate. VOK xi

A truly wonderful thing about the TSK vision is that we can apply inquiry to the vision itself. Nothing is out of bounds. So let us question this last quotation. How can we ask "with the heart, the senses, and the body as well as the mind?"

The only possible answer to this question comes from trying this kind of inquiry out. That is what the exercises in this program aim to do. They explore the prospects for inquiry in the same way we might explore a new city in a foreign country, throwing away the guidebooks (but it would probably help to study them first!) and setting out on our own.

The nature of this exploration, this way of conducting inquiry, is inherently playful, a quality reflected in much of Rinpoche's writing on TSK:

> Why speak of space, time, and knowledge at all? Because the transition from ordinary knowledge requires a point of departure. As this vision unfolds, this requirement loses its hold. Each point opens up, becoming a zero point. We could say that the vision is ultimately 'pointless', but we could also say with equal conviction that each point is pointed out as the Great Point at which Great Time and Great Space meet with Great Knowledge. Both characterizations are points of entry. We speak for the time being only, relying on words as symbols or gestures, ways of bringing knowledge into time and space. VOK xii

How do you react to this approach? Does the paragraph just quoted seem too much like word games? Try treating it playfully yourself. Read the paragraph a few times, chewing it in your mouth with appreciation. Perhaps set it aside and come back to it later. Where does it lead you?

There is no need at this point to define precisely what TSK inquiry is. We know it is not conceptual insights, but it is also not "direct" experience. See for example, VOK xv, which rejects a reduction of TSK (in its Time aspect) to the 1970's slogan "Be here now" (compare TSK 85). Our not knowing, as VOK xvii points out, is itself the opening we need for knowledge to emerge. This theme—the value of starting from not-knowing—comes up repeatedly in TSK inquiry. For one example, see the discussion of confusion and not-knowing at VOK xvii; for another, see LOK xxv.

The Fruits of Inquiry

What are the benefits of TSK inquiry, of exploring the TSK vision? The answer can be hard to put into words, because TSK rejects the "improvement orientation" that tells us we have to make progress along a path, have to change something. "Since nothing is missing, why look for something to fix or complete?" VOK xii

This may sound like pop psychology: "Everything is perfect just the way it is." "It's all good." But if we understand that we are turning away from an improvement orientation, the meaning is much more specific. We might think of it this way: There are two kinds of knowledge. Our standard education emphasizes one of these: accumulating data, concepts, and experience, until we come to understand something we did not understand previously. But TSK is about a different kind of knowledge, which might be described as the whole of what we know, or even better as the whole of the ongoing activity of knowing/being known. That knowledge is not cumulative; it does not "add up." It is the *quality* of our knowledge that changes, not the amount of it.

In this sense, then, there is nothing to fix or complete, because the whole is always the whole. We could say that this second kind of knowledge—knowledge of the whole, sometimes referred to in the TSK books as "global knowledge"—is more available or less available, but even that is not quite right, because that way of putting it suggests a fixed body of knowledge. The TSK books talks often about the body of knowledge, but this body is not fixed. It is a living, breathing body of knowledge, brought alive in the act of knowing.

So what changes as we engage the TSK vision? Perhaps we can say simply that through TSK inquiry, we no longer live as we have always lived, no longer know as we have always known. But how can we put this into words, even to ourselves? VOK xv speaks of the "immediate presence" of knowledge, and perhaps that is enough.

Exploring the Limits

Inquiry is central to the TSK vision, but inquiry must be guided. With this in mind, it may help to think in terms of three aspects of the vision. The first is inquiry itself, the active dynamic that brings the vision alive. The second is analysis and exploration of the limits we experience in our usual way of being and acting: inquiry focused in a specific direction. And the third is an evocation and activation of ways of being that transcend these limits. Here too inquiry is focused in a specific direction, but it also seems fair to say that with this third aspect the nature of inquiry itself changes, for otherwise, how could transcendence be possible?

In Part One of this program, the focus of inquiry is mostly on limits: on learning to see more clearly how we have closed down experience, filled up space, shut down time, and put a halt to active knowing. But the distinction is not hard and fast. All three of these themes can be found in just about every chapter of the TSK books.

In the readings for this unit, the theme of limits comes up most clearly with respect to time, and a link is made between the dynamic of inquiry and the dynamic of time. See VOK xv: "Yet once we activate inquiry, we discover another dimension of time," a dimension that allows for a space filled with knowledge. This link between inquiry and the temporal dynamic is something we will return to later.

The readings for this unit are unusual in drawing on all six books. In part this is to encourage you to work with the CD, and in part it is so that you can get a sense of each of the books in the TSK series. But it is also because the readings complement each other, returning to the same themes from different perspectives.

For the most part, the readings speak for themselves. Be sure to look at DTS xvi–xx, which investigates what TSK is *not*. Another very useful passage is the first two paragraphs on KTS xvii. Notice the distinction between experience and the content of experience ("what is given" vs. "the way it is given"), a distinction that you can explore directly in Exercise A, below. Notice also the very important distinction starting at the bottom of KTS xvii between space, time, and knowledge as aspects of our experience and Time, Space, and Knowledge "as interacting dimensions of our being."

The idea that we can "let old patterns of action, perception, and understanding drop away, *without immediately installing new ones in their place,*" is central to TSK inquiry. You may not have a very clear sense of what it would be like to operate without patterns, but just asking the question invites you to notice the patterns that operate in your life, a very important move in living the TSK vision.

If we dropped the patterns that run our lives, would we be free? Before we can answer that question, we need to develop an understanding of what it means to be free from the perspective the TSK vision.

Speaking in a very preliminary way, our usual notion of freedom is founded on the notion of self. It is the self that wants to be free; that insists on the right to be free. Freedom in this sense is a psychological concept, and it remains psychological even when the more sophisticated claim is made that the self must learn to free itself of self-imposed limits and patterns.

The TSK vision takes a far more radical approach, setting aside the focus on the self in favor of Time, Space, and Knowledge as facets of being. That is why it can be said to introduce a new *vision* of reality. There is a natural tendency to reduce the TSK vision to a personal kind of undertaking, but actually TSK does not operate on a personal level at all. Take a look at KTS xviii–xix for some thoughts on these two different levels and how to work with them.

Exercises

The first exercise for this program is not found in any of the TSK books. It was introduced by Rinpoche in the programs he led on TSK in the 1970s, and has often been used on retreats since then. We can call it Exercise A.

Exercise A: Expanding and Condensing

Settle into your experience, and then begin to expand or condense it. In doing so, keep in mind that experience is made up of many aspects: thoughts, sensations, feelings, identifications, judgments, perceptions, the relationships among these elements, and other features that may be too subtle to name. You can expand or condense each or any of these, or you can expand or condense the whole of your experience. For instance, you could expand the sense of distance that separates you from whatever you see in front of you, or you could condense your sense of self. You could expand a sensation you feel in your shoulder, or you could condense the feeling of time passing, or the vague feeling of anxiety that the passing of time evokes. Do the exercise in any rhythm you like, fast or slow, regular or random.

Comment to Exercise A

In doing this exercise, feel free to play with experience. You can loosen limits, borders, distances, and perspectives. It is like pulling on a stretchy fabric, which can take any shape at all or even turn into something else. To warm up, it might be helpful to focus first on sensations, then move on to thoughts, memories, and fantasies, and then go in whatever direction seems most inviting. There are really no limits; for example, you can expand or condense the activity of expanding, expand or condense the sense of ownership, expand or condense the feeling of not knowing what you are doing, and so on. Notice that there are at least two different levels to many experiences: the content of the experience, and the experiencing of the experience. You can expand or condense at both levels.

Most people find it easier and more enjoyable to expand than to condense. Explore this for yourself, and see why this might be so, if in fact that is your experience. If expanding is easier, you can focus more there, but don't forget about condensing entirely. Keep in mind that "condensing" is not the same as "contracting." It is not just a matter of making something smaller, but of making something more condensed. Condensing is related to concentration, and to adding more dimensions to experience.

One nice thing about the practice of Expanding and Condensing is that you can do it at any time: while waiting in traffic, while reading, or while having a conversation. While you are working with this unit, try doing Expanding and Condensing not only during the time you set aside for TSK practice, but also in your daily activities. Don't forget to make notes in your journal about what you find.

Since this is the first exercise we are working with, it may be worth making a few observations about TSK exercises. First of all, the exercises are not meant to produce a specific result. They are explorations and inquiries, not techniques. Second, don't be surprised if the same exercise gives very different experiences each time you do it. By the same token, you can expect that sometimes an exercise will be very interesting, while other times it will be very boring. Usually you can find a way to bring this boredom into the practice of the exercise, and then things may get interesting all over again.

Finally, the TSK exercises ask us to investigate in unexpected and unfamiliar ways. This unfamiliarity is an important part of doing the exercise. In fact, if an exercise starts to feel familiar and routine, and especially if it always unfolds in the same way, this is an indication that it is time to take some new approach.

Because the exercises challenge familiar ways of seeing and doing, it is natural to be confused about how to do them. Sometimes you may find yourself with no idea at all how to do an exercise. That doesn't matter. Do it anyway!

Exercise TSK 24: Marriage of Sound and Breath

While Exercise A invites a more open and playful approach to space, this exercise is a pathway into the dynamic of time. As the instructions point out, it is important to focus on the quality of the sounds you hear, not their significance. This relates again to the distinction between experience and the content of experience, a theme that comes up often in the TSK vision.

The instructions to this exercise speak in terms of going beneath "surface partitionings," and also ask you to focus on the energy in the throat center. You might ask: If the energy of the throat helps dissolve or loosen partitions, how are these partitions formed in the first place? Are they associated with another energy center in the body? Or is it tightness in the throat that helps set these partitions (or limits) up and keep them in play?

The comment to this exercise in the TSK book suggests doing it three hours a day. Don't worry if you can't find the time for this kind of intensive practice. You will still find the practice beneficial.

Unit Two
The Ownership of Knowledge

LOK pp. xxxix–l; 33-40; SDTS pp. xiii–xviii, 3–14
Exercises: TSK Ex. 7, 23
Supplementary Exercises: LOK Ex. xxvii; LOK Ex. 5

The previous unit suggested the TSK vision is not psychological in its orientation. Instead, it questions the psychology of the self at a deep level, turning away from a personal actor at the center of experience. This unit explores that turn.

A good way into the central theme for this unit comes from a passage in the readings for Unit One:

> One way to investigate the *human significance* of Time, Space, and Knowledge is through an inquiry into balance. As individuals, we are often out of balance, whether on the psychological level, the physical level, or in our relationship with external circumstances. An experiential inquiry can open Space, energize Time, and make Knowledge available, restoring a balance that brings harmony to the way we live our lives.

> From within the TSK vision, however, this view is only a first-level approximation of the true potential for balance. The language we use inevitably directs us toward the individual actor at the center of any experience, but the balance attained by the 'actor-self' is a 'personalized' version of the interplay among Space, Time, and Knowledge. The self singles out *as its possessions* emotions, self-image, attachments, relationships, and a sense of personal freedom. It sets in motion a momentum that heads at once toward imbalance.
>
> The style of inquiry introduced here counteracts the tendency toward 'singling out', thus signaling a return to balance. With no positions or possessions, such inquiry is committed only to knowledge, and this commitment, *because it remains available as a subject for inquiry*, is in an important sense self-correcting.
>
> Complete commitment to inquiry can be maintained only through the activation of a subtle, probing intelligence. The self tends to *adopt* a position that it describes to itself as being *free from* positions. The resulting confusion and self-deception are similar to a drunkard's conviction that the most ordinary thoughts are profound and revolutionary insights.
> LOK xxix

The theme of 'no-ownership knowledge' emerges very clearly in these paragraphs. Incidentally, notice the play on words when Rinpoche writes that the TSK commitment to inquiry is "self-correcting." You might want to reflect on a similar claim made in Western science, which could actually be understood as involving the same play on words. What is similar in these two approaches? What is different?

The assigned reading from the front matter for LOK addresses the theme of no-ownership knowledge powerfully on xli–xlii, and in a more light and engaging way starting on xliii. The idea of not attacking the self directly (p. xli) is important to keep in mind. We will want to look at it again later in the program, when we investigate the self in more detail. Take a look especially at the discussion on LOK, xlvii of the "painfully restricted" kinds of knowledge the self is able to own.

Ch. 5 of LOK develops this insight with great power. In the years since I first studied *Love of Knowledge*, I have come to appreciate this chapter more and more. Short though it is, it contains a critique of the whole Western understanding of what counts as knowledge, and of the fact/value dichotomy in Western thought. Whole books have been written on these subjects that do not get at its core as powerfully as this short chapter. Especially interesting is the discussion on LOK, 37, which explores why it is that insights do not affect us in our being, and that even the most transformative experience will soon fade. If you understand this point thoroughly, it will be a great help in the rest of the program, and of course in your life as well. We are all looking for change and all hope for transformation, but if we do not challenge the usual structures of knowledge, and the usual structures of space and time that support them, these efforts will be self-defeating.

Before going on to discuss the reading from SDTS, it might be good to ask how we are supposed to approach these readings. When Rinpoche writes that we get into trouble when we assume that the self is the owner of knowledge, he seems to put us in a box. On the one hand, there is no point in just accepting this statement as true. Doing that is just a way of putting the claim that it makes at a distance. We say, "The self is not the owner of knowledge." But under our breath, we say, "I know that," laying claim to ownership!

This difficulty arises because we think that knowledge operates on the level of ideas. Rinpoche has addressed this problem in another book, *Knowledge of Freedom*:

> What is happening in our minds as we read and reflect on the ideas written here? . . . Are we stepping

> out from our accustomed ways of knowing, or are ideas only pointing to ideas . . . ? As we read, are we picking and choosing what fits with what we already understand? . . . Whatever we say to ourselves, aren't we simply repeating the same patterns? . . . Even the finest thoughts we have—thoughts of awakened mind, of awareness, of enlightenment—are names and concepts interpreted and judged by the mind.
>
> KoF 332

The TSK vision recognizes this contradiction and offers a way to resolve it. Its approach is to go to the deepest levels of our experience, before ideas, before ownership, before the patterns of mind.

The readings from SDTS again address the issue of self as owner of experience. The vocabulary here is unique to SDTS, which re-visions reality in terms of a knowledgeable space-time geometry. Don't worry too much about the discussion of the zero-point and 16. They will come up again much later in this program. For now, it is useful just to work with thinking in unfamiliar ways and with unfamiliar vocabulary. Later on we will look in considerable detail at the importance and value of not-knowing as the gateway to a world in which we can leave our usual ways of knowing behind. The readings for this unit remind us how stuck our usual ways of thinking and knowing are, so it helps to try on some new ways of thinking and knowing as you read, not insisting on understanding exactly what you are doing. For some help thinking this through, see topics 21, 63, and 64 of the TSK Guide.

Exercises

Ex. TSK 7, Body-Mind-Thought Interplay, is similar to mindfulness practices found in different meditation traditions, but it takes on a special flavor in the context of the TSK vision. Here it is used as an invitation to let experience arise without any claims to ownership or presupposed identifications.

This invitation is one we can welcome gladly and make use of in very practical terms. Sometimes when you do a TSK exercise, it can be difficult to settle into it. Other thoughts come up, driven by concerns and worries, feelings of excitement or desire, or a fundamental restlessness. Instead of regarding such experiences as an obstacle to be overcome before you can start practice, take them as an opportunity to practice Body-Mind-Thought Interplay. Gradually experience will settle down, and you can turn to the main practice.

Ex. LOK 5 looks very directly at the claim of the self to stand outside experience, and suggests that this claim can be challenged. You may well be frustrated or uncertain how to do this practice, but this should not stop you from trying. This is often true for TSK exercises: a initial tendency to be confused or unclear about how to proceed, which may with a little patience open in unexpected directions.

One of the two supplementary exercises, TSK Ex. 23, Going without Going, is a favorite for TSK practitioners, but many people find it easier to do in a retreat or other group setting. It is a natural for this unit, since "going without going" leads directly to "going without a goer;" that is, to a doing in which the self is not a doer. If you do try the exercise and find yourself frustrated, just ask, "Who is frustrated? . . . Why?"

The other supplementary exercise consists of a sentence at LOK xxvii: "The inquiry here investigates the fundamental process by which human beings turn perceptions into judgments, judgments into patterns, and patterns into fixed positions" (compare LOK Ex. 21). The practice is to see if you can distinguish these four stages—perceptions, judgments, patterns, and fixed positions. Here too, the idea is just to see what happens when you try. Taking notes in your journal afterward may help you make sense out of the experience you have in doing the exercise. If you conclude that it is hard to tell these four phases of experience apart, what does this tell you about our capacity to know the mind? These distinctions may take on added significance as the program continues.

Unit Three

Three Levels of Time, Space, and Knowledge

LOK pp. xix–xxiv; SDTS pp. xxix–xxxvii;
TSK pp. 111–114; LOK pp. 365–68
Exercises: TSK Ex. 30; DTS Ex. 1
Supplementary Exercises: TSK Ex. 34; TSK Ex. 16

This unit completes the introduction to the program. The focus is on the three levels of space, time, and knowledge, a central (but mostly implicit) theme for the rest of the program, and the organizing principle for the TSK Guide.

The key reading for this week is from the front matter of SDTS. Notice especially the discussion on xxxi on how the second-level approach to time, space, and knowledge contrasts with the 'improvement orientation' that we ordinarily apply when we set out to change our situation:

> This way of seeing offers something new—not because it is superior to prevailing systems of knowledge or

> belief, but because it steps outside the framework within which such systems arise and develop their perspectives. That is why TSK is not concerned with praise and blame, hope and fear, or right and wrong. It can be naturally accommodating of every alternative, because it has no position to defend or point to make. In fact, a second-level TSK perspective makes it easy to transform or transcend such dichotomies, for it reveals the mechanisms of their arising.

As a kind of informal exercise, see if you can trace how often the dichotomies of praise and blame, hope and fear, and right or wrong operate in your life. These patterns are both subtle and pervasive, and you probably won't see how often they work. But when you do, ask if there could be another way to live. More precisely, ask if you can see the mechanism of their arising. That is one way to explore how a second-level view could have a transforming impact.

The reading from TSK is a concluding summary of a much longer discussion of the three levels of space. There is a tendency to think of first-level space (or first-level time and first-level knowledge) as being equivalent to ordinary experience. But the reading qualifies this assumption in an important way at TSK 111:

> Space, level one, is precisely our ordinary realm, seen in the light of further possibilities for human fulfillment.

How can we cultivate an appreciation for such further possibilities? The exercises for this unit have been chosen with that aim in mind. It is not so much a question of having special experiences; rather, the point is to see what it might be like to sense that our experience can be different, can go to a different level. As the comment to DTS Ex. 1 says:

> The more you practice the exercise, the more you will recognize that the aim is not to find new knowledge or to succeed in doing what is asked. This way of structuring the exercise will only confirm the

> view that the knowledge we ordinarily operate with is incomplete. Here the challenge goes deeper. If you go closely into full awareness without preserving a realm of messages or concepts, you can conduct openness.

Notice the link made in the comment on this exercise between openness and not-knowing.

Although it is presented as a supplementary exercise, TSK Ex. 16 is a very direct invitation to explore different levels of time, space, and knowledge. See if you can do it a least once or twice before moving on to the next unit. Read the commentary to this exercise lightly, since it refers to concepts that have not yet been introduced in this program.

As you do the readings for this unit, look for places that seem to conflict with one another, or that raise questions you wish were answered more explicitly. The readings from near the end of LOK, which do not specifically speak of levels, but certainly seem to evoke three different approaches to experience, may raise these kinds of issues. If the readings seem abstract, rely more on the exercises.

There is a tendency for TSK students to imagine that they have 'had' a level-three experience when they are still firmly rooted in 'lower-level' ways of acting and being. In the end, though, trying to assign your experiences or insights to different levels is not that helpful. Just be open to experience and ready to challenge limiting structures, and your experience will naturally expand and deepen.

Unit Four
Space and Embodiment

TSK pp. 3–18
Exercises: TSK Ex. 1–3; TSK Ex. 4–6

This unit starts a new part of the program. For the next three units, we will review the space-time-knowledge interplay. This first week, we focus on space, reading from the first TSK book. The aim is to get clear on the significance of space, and on the way that we tend to take space for granted, because we are so accustomed to focusing instead on the things that *appear in* space.

A key insight in beginning to focus in on space has to do with focal settings, discussed at pp. 4–6 of the readings. There is a helpful related passage in SDTS that gives some added dimensions to this concept, and also helps link it to the exercises assigned for this unit:

> When we enter a well-designed house, we experience various qualities rooted in the relationships of form, space, color, movement, and light. Even when we are not consciously aware of these reactions, our

> response to being in the house springs from a very quick and efficient 'sizing up' of the relationship of different parts to one another, and ultimately to the whole of the structure.
>
> Suppose we left the house and passed through its yard into the streets of the town. We would now be perceiving a new and wider array of relationships: the relationship of house to house, block to block, and so on. As we reached the edge of the town, we would move on to yet wider relationships: town to countryside, countryside to seashore, continent to continent.
>
> There is no pre-established limit to the scope of such an investigation, whether we move out into the larger universe or explore the complexity of the microscopic and subatomic realms. Whether we were examining a water molecule, a work of abstract art, or a supernova, certain underlying principles would apply. These principles include the perception of difference, distinction, order, structure, and scale, all directly experienced as a certain living and dynamic quality that gives rise at another level to our judgments of like and dislike. SDTS 38–39

It may also be useful to think of the different focal settings that apply when we are driving through the countryside in a car. If we focus in terms of the car, we are sitting still. If we focus in terms of the countryside, we are moving along at a good pace. Again, think of the experience of sitting in a train station next to another stopped train. If the other train starts to move, it can be very disorienting, because for a moment you cannot tell whether you or the other train has set into motion. This an example of focal setting in action.

Similar considerations apply if you think of the space we inhabit when we are busy thinking thoughts. Perhaps it seems strange to extend the idea of 'space', which is usually thought of as physical, to the mental realm, but the sense of focal setting allows us at least to consider this possibility. If we say (in a preliminary, first-level sense) that the space we perceive/encounter/interact with is a function of our interest or concerns, then there is no reason to limit 'space' to the physical domain.

In TSK, 3 (sky as space), and again in TSK, 12, the notion is introduced that 'ordinary' space could be conceived of as a 'representative' of space in a more encompassing sense. Reflect on how this connects to the definition of space, level one, from the last unit: our ordinary realm, seen in the light of further possibilities for human fulfillment. This also ties back in to the discussion in earlier readings of how space as we ordinarily encounter it is crowded and limited.

Even though the reading for this week is the very first chapter of the very first TSK book, it is quite challenging. There are three important and difficult sentences in the chapter that are worth reflecting on:

> If we employ new 'focal settings' and see the way they work, we can come to *an overall understanding which is itself a kind of space.* TSK 5
>
> One very useful and positive consequence of [the] ambiguity [between higher- and lower-order perspectives] is that even conventional dichotomies . . . preserve, in a symbolic or 'inspired' sense, expressions of a higher order unity. TSK 9
>
> Space is projecting space into space. TSK 10

Of these statements, the first is perhaps the most puzzling. It seems to point toward the relationship between space and knowingness, a topic that we will look at more carefully later in the program. The concept of a focal setting is helpful here, a reminder that having a view is related to discovering a particular kind of space to be available. In general, the relation between 'seeing' in the physical sense and 'seeing ' in the mental sense is very suggestive and wholly in accord with what is said in the second of the three quoted sentences.

This second sentence is actually fairly clear when read in context; if you want to explore it, see the references in the TSK Guide under Topic 21 (Symbols). It relates back to the idea already mentioned that lower-level space can be seen as a representative of a higher-level space.

The third sentence can be considered at several levels. In first-level terms, it is simply a rather unusual formulation of something that is hard to deny: Since objects 'occupy' space, whenever anything happens in conventional terms, space is the ground of that happening. We usually say that 'things' move or act, but all the moving and acting depends on space, and that is one way to understand the claim that space is projecting space into space. On a second level, the sentence grows directly out of the second of the three sentences: What looks like a solid object on the lower level can also be seen as the (symbolic) expression of a higher-level space. On the third level, the characteristics that we consider essential to reality (and that make this statement seem so paradoxical) are themselves understood as aspects of space, a point that relates to the first of the three quoted sentences. For more on this third sentence, you might look at an essay I wrote on it in *A New Way of Being* (Dharma Publishing, 2004), pp. 159–74.

Exercises

There are actually six exercises assigned for this week, but it makes reasonably good sense to think of them as two exercises, each of which has three parts. The first exercise (Ex. 1–3) could be called "Space and Embodiment," while the second could be called "Bringing Knowledge into Space." Before you try the exercises (generally known collectively in TSK circles as the Giant Body Exercises), take a look at the discussion on page 7 about the difficulty of looking at space directly. The exercises deal with that difficulty by letting us experience space at different levels, or in connection with different objects and at different focal settings. The suggestion that emerges from the series of exercises is that if we go far enough, objects at any level resolve into space. You may also want to look at the "thought experiment" that opens DTS (DTS 3–5).

"Space and Embodiment" operates on the ordinary level, asking us simply to develop an appreciation for the space that makes up so much (if not all) of conventional reality. The opening of Chapter 2 in the TSK book, where these exercises are introduced, suggests that it makes sense to challenge our usual conception of the body, because "our bodies are a central focus for our experiences as human beings." (TSK 21) However, the Giant Body approach to doing this comes as something of a surprise, because it does not rely at all on our experience of our bodies; instead, it challenges our *conceptual* understanding of what makes up the body. In this sense too the body is a useful focus for inquiry, because we are all generally familiar with the various layers that make it up, from organs to subatomic particles. On this point, you might want to look at TSK 40–42, assigned for Unit Ten.

Many readers of TSK have great difficulty with the Giant Body exercises, because they call for an ability to visualize that goes well beyond what most people believe they are capable of. Just do what you can, and gradually some clarity will emerge. While the text (TSK 41) does say that accuracy is desirable, keep in mind that the purpose of doing the exercises is to develop an appreciation for space. It is a little like being able to appreciate the beauty of a sunset by focusing on the clouds, which is something we do naturally. So do not be overly concerned if you can visualize only in general terms or can only get clear on this or that specific.

Part of the difficulty with doing the Giant Body exercises comes from the dichotomy that identifies a self that is doing the visualization and the body that is the object of the visualization. Instead, try letting the 'object' of visualization appear in space, in accord with an intentionality that does not exactly belong to the self, but is instead inherent in the situation as a whole. The commentary to TSK Ex. 3 helps clarify this point.

In any case, it would be a mistake to focus too strongly on the specifics of doing the visualization. The purpose is not to get precise and clear images, but to see how form dissolves into space (see, e.g., the commentary to TSK Ex. 2). If you find that the visualization comes easily, open up the inevitable feelings of accomplishment and self-satisfaction. This way of 'opening up' is addressed specifically in later exercises from the TSK book, starting with TSK Ex. 8–9 and the discussion of the 'coming out' tendency of the self), but here you can do it more informally. You might also apply Ex. A, Expanding and Condensing, from Unit One.

The focus on visualization in the Giant Body exercises prepares the way for a more subtle set of practices later on. Unlike 'real' things as we ordinarily conceive of them, visualized objects are by definition 'charged' with knowingness; that is (again, in first-level terms), they are mental constructs. This means that when we visualize, we are bringing space and knowingness into contact. Of course, we continue to regard the visualized object as only imaginary, only subjective. But later, we will challenge the solidity of the one who is visualizing, and then this simple dichotomy between real and imaginary proves unworkable. By visualizing 'real' objects such as the body, we prepare the way for 'visualizing' the self (a process already initiated in the practice of LOK Ex. 5 and TSK Ex. 30).

As for the second set of three exercises assigned for this week (Bringing Knowledge into Space), TSK Ex. 5 makes a direct connection for the first time between (visualized) objects in space and a purely mental quality (ecstasy). You might view this as a bridge between space in the lower-level sense and a higher-level knowingness. The bridge goes by way of the feelings, and especially through bringing a sense of aliveness to what appears in space, which is exactly what this set of exercises emphasizes.

In this second set of exercises, the earlier emphasis on 'penetrating' one level to reach the next gives way to a sense of releasing or opening. Both these movements are important (compare the rhythm of expanding and condensing), and both are a part of TSK inquiry. 'Penetrating' could be thought of as linked to time (in a first-level sense): a direct, linear movement (see TSK 141). Releasing is more linked to space. However, at the second level (where the self as such has already been penetrated to some extent) it may be that time is more associated with releasing (cf. TSK 128–32), while space engages a dynamic that may be linked to penetration.

Unit Five
Past Present Future

TSK pp. 117–41
Exercises: TSK Ex. 18-19; TSK Ex. 20-21

In this unit we turn to the theme of time, focusing in particular on the past-present-future structure of time. The reading looks at the structure of time in relation to events that occur 'in' time. We are just at a preliminary stage, and what matters here is to recognize that when we get clear on the usual model of time and see its shortcomings, we are ready to ask whether an alternative may be available. As to the nature of that alternative, the reading suggests (in parallel to the treatment of space, which we looked at in the last unit) "that objects, their states, and their interrelationships *may somehow be a function of 'time' itself.*" TSK 120

Most people would agree that time in some sense *flows* from past to present to future, or perhaps from future to present and on into the past. Yet the reading suggests that due to our focus on 'things' and identities, we do not ordinarily connect with this flow:

> The first stage of relating to time is devoid of any direct experience of the 'lower time' which it is said to involve. TSK 120

In the same discussion, Rinpoche writes that we ignore time, and that it is because of this ignoring, or ignorance, that time can seem so oppressive. You should reflect on this statement with care, instead of just accepting it. What is your relationship to time? Do you experience it as an outside force? Do you experience time at all? Is experiencing events 'in' time the same as experiencing time? When you are in the middle of an event (for instance, reading this paragraph), do you have a sense of time passing? Or is that just a conclusion you reach when you reflect back on what has happened?

How are past and present and future linked? If it is true that we are ordinarily cut off from the flow of time, then asking this question will require us to contact second-level time, and even third-level time (Great Time). For an introduction to what this would mean, take a look at TSK 98–101. The discussion here takes place in the context of a commentary on TSK Ex. 15, which we have not looked at, but you should be able to follow what is being said without too much difficulty. Of course, we are not yet ready to say much about what Great Time 'is' or how it operates, but the reading does not ask that we do so. For more on these themes, see the readings for topics 39–41 and 48 in the TSK Guide.

Once we have recognized the possibility of other ways of relating to time (recall the definition of first-level space: "our ordinary realm, *seen in the light of further possibilities* for human fulfillment"), we can begin to investigate the damage done to us by our lower-level view of time. Time is the 'stuff' of our lives, so it should be easy for you to explore in your own experience and at a psychological level how we are related to time. Of course, the TSK vision is not itself psychological, but the wants and needs, concerns and wishes that we all experience are the starting point for inquiry. They pervade our experience and shape our reality, and we have to be willing to bring them into play if we want TSK inquiry to move beyond the level of concept and theory. See especially TSK 128–32. TSK Ex. 18b, assigned for this week, offers a direct and practical way to carry out this kind of ongoing inquiry.

The reading for this unit actually distinguishes two aspects of lower-level linear time. One is its characteristic as the neutral grid for experience: one second after another after another. The other is time as a force bearing down on us from outside. The discussion at TSK 136–41, aims to show how these two aspects—the subjective and objective components of lower-level time—are related. In reflecting on what is said here, it might help to review the readings for Unit Two, which suggest that this split into the subjective and objective realms can itself be challenged.

The TSK vision does not aim to undermine other approaches to transformation, but it does sometimes point to their limitations. The reading contains a good example:

> We may become alarmed by the situation of being enmeshed in 'lower time', with its appearance of particular times and intervals, and may even attempt to break free of it through practicing meditation or through religious disciplines. But since we generally structure such efforts in terms of the familiar character and logic of things, we may not get very far.
> TSK 139

If you have a background in a meditative or spiritual discipline, you may want to reflect on this claim. Does the TSK perspective suggest a way of deepening your practice of these other disciplines? Would a different understanding of time change the way you approach such teachings?

Exercises

As with the last unit, the exercises assigned for this unit actually combine several TSK exercises. Ex. 18–19 taken together could be called "Threefold Time." Ex. 20–21 together could be called "Temporal Reversals." The first set of exercises invite you to get familiar with the temporal structure, while the second set asks you to vary and challenge that structure (much as Bringing Knowledge into Space challenges the usual understanding of space. TSK 125 invites us to "work directly with 'time' and see *why* it 'goes'," and this can be a good starting point for the exercises for this unit.

The exercises for this week go directly to issues of self-identity. On the one hand, this tends to make them easily accessible and rewarding to practice. On the other hand, you will probably 'hold out' or reserve certain parts of your experience from the exercise, because it is too closely tied to your self-image and identity. If you want to explore this aspect of the exercises (especially TSK Ex. 19), you could try combining them with LOK Ex. 5, assigned earlier.

Temporal Reversals is a more challenging exercise than Threefold Time, but at least the intention is clear. As a way to go deeper into the exercise, consider this passage from the commentary to TSK Ex. 21:

> It is precisely this subtle or overt directedness that impoverishes the present and gives rise to real, sequential time—guaranteeing the emergence of further impoverished but 'real' moments. TSK 176

Can you see in your experience how this connection works? Do you get a sense from doing the exercises that it could be different? Do you agree that the exercises "contribute to a transformation of 'real' time?"

If your answer to this last question is yes, be careful that you are not agreeing too easily. We are much more tightly bound to linear time, which is so central to our own identity, than you may imagine. To explore this point, you might try practicing Temporal Reversals with a special focus on the past-present-future structure that supports the identity of the self as the one doing the practice.

Unit Six
Opening Perspectives

DTS pp. xxv–xliv; SDTS xix–xxix
Exercises: TSK Ex. 17, LOK Ex. 38
Supplementary Exercise: TSK Ex. 34

In the last several units, we have focused on the first TSK book. In this unit, we go back into the front matter of two of the later books, reading material that focuses on the special sense of 'knowledge' that the TSK vision seeks to invoke. The fact that we are still reading in the front matter is a reminder that we have only begun to scratch the surface of the TSK vision. At the same time, the readings help put the previous work with time and space into a broader perspective. In a sense, they disclose and make sense of the path upon which we have already embarked.

The focus of the DTS reading is on developing a methodology for knowing the mind. Ask yourself how this relates to the work with time and space that we did in the last two units. For instance, a focus on 'mind' sounds subjective, whereas time and space are in some sense 'objective'. But by now we know enough to question this distinction. In practicing Temporal Reversals in the last unit, we were 'only' engaging subjective experience (see the commentary at TSK 176), yet we were also looking at time itself. Could the same be true in reverse when we look at the mind?

As with space and time, a living inquiry into knowledge demands that we get clear on our ordinary understanding of knowledge. Many people are resistant to this kind of inquiry, and this resistance is interesting in itself.

In this culture we learn from an early age that time and space may be different from what we think. Not only does science tell us that time is relative and 'empty' space is far from empty, but advances in technology are constantly making us rethink the limits of time and space, which seem to change from generation to generation, or even year to year. When it comes to knowledge, however, things look very different. The same science that tells us that time and space are open to inquiry insists that the nature of knowledge is fixed and indisputable, defined by science in terms of what can measured and explained and by philosophy in terms of what can be put in the form of well-formed propositions.

The result is to put knowledge itself off-limits to inquiry, which makes the very word "knowledge" sound boring and uninviting. In fact, some thinkers define knowledge as inherently limited. They invite us to look instead for an active 'knowing' that does not let itself be *reduced* to mere 'knowledge'.

The TSK vision is sympathetic to this approach, but it is not ready to give up on ordinary knowledge. Just as linear time can be the gateway to the dynamic of time and empty space can open into the freedom of Great Space, so ordinary knowledge, with its subject-object structure, its propositions, proofs, and presuppositions, can invite Great Knowledge.

To bring about this transformation, we need to investigate conventional knowledge. The readings from DTS introduce a number of important and related themes that focus in on the structures of conventional knowledge: the importance of categories and interpretations (xxxiv); the inner structure of explanation (xxxvi); and the interplay of thought, models, and limits (xxxviii). Each of these points is linked to conventional understandings of time or space, reminding us of how these three factors of being interact. For instance, consider this brief but profound discussion of explanations (xxxvi):

> Why do we find that we cannot act in accord with our own convictions and values? It is easy enough to offer explanations: our psychology, our conditioning, our physical being. But explanations mirror the temporal structures of the conduct they attempt to explain: They track one level of sequencing with another.

For more on the interplay between our understanding of time and space and our ability to inquire into knowledge, take a look at DTS xxxiii.

In contrast to the reading from DTS, the reading from SDTS goes more toward offering an alternative to the structures of conventional knowing, focusing strongly on the nature of inquiry. In this context, it introduces the idea of logos (p. xxviii). We will return to this important conceptual tool later; if you want to explore further, see the TSK Guide, Topic 65. Here is a very helpful definition of logos:

> The governing understanding active within a temporal order. The 'logos' operates in accord with a specific and characteristic logic, establishing possibilities and defining limits. On a lower level, the 'logos' is invisible, operating as 'the (temporal) way things are'. With more knowledge, the focal setting shifts, and the 'logos' becomes accessible as a structure to be investigated in terms of prevailing 'read-outs'. As investigation continues, the distinction between 'logos' as knowledge and 'logos' as the prevailing order of time and space gives way to a more comprehensive 'knowingness'. LOK 415

The reading from DTS offers many important themes: learning to look with knowledge eyes (xxvi); the distinction between possessing and embodying knowledge (xxvii); knowledge as a natural part of being (xxviii). Do not let the eye and mind glide over these ideas: Do you know what is being described? Have you begun to shift your own knowing in that direction? Do you know something about how to do so?

Ultimately, the DTS reading proposes a specific methodology for TSK inquiry, starting at the middle of DTS xxx. You could treat this passage, especially the second full paragraph on DTS xxxi, as an exercise. In any case, reflect on the link that Rinpoche makes here between knowing and being:

> Such possibilities will come alive when we truly realize that the mind is not a tool that we can operate independent of who we are. For mind to function differently, opening differently to knowledge, we must learn a different way of being.

Exercises

TSK Ex. 17, The Object and its Glow, is introduced in the section of TSK on time. As the commentary to the exercise suggests, it could equally well be seen as a time exercise, a space exercise, or a knowledge exercise. Here we are doing it in terms of knowledge. The discussion of the object and its glow that the exercise refers to and presupposes is found in TSK, 171.

As we have already seen repeatedly, in the usual understanding, knowledge is something that a knower knows. Here, as with Subject-Object Reversal, the point is to de-emphasize the self as knower. Notice that three elements are identified as operating in the conventional model for knowing: the subject, the object, and the distance between them. Of course, the distance being referred to may not be a physical distance. For instance, the object you have in mind may be a remembered event or a model for experience.

LOK Ex. 38, Embodied Energy, is also a challenge to the usual subject-object structure of knowing. Notice that you are not focused on an object external to yourself, but on your own feelings, sensations, and other positionings. At some level, however, this distinction breaks down: When you are aware of an object, you are aware in an embodied, whole-being way, and when you focus on your own inner experience, you are at least in part making that experience an object for inquiry. Actually the relationships are far more subtle than these summary statements suggest, as you will discover when you do the exercise.

Unit Seven

Space of Thoughts

TSK pp. 49–68, KTS pp. 163–67
Exercises: TSK Ex. 11; DTS Ex. 2

In the last three units, we went through the space-time-knowledge cycle for the first time. Now we will do it again.

With respect to space, this second cycle has a specific purpose: to introduce the notion of mental space, or the space of thoughts. This is important to the TSK vision as a whole, because this expanded view of space makes it easier to introduce the idea that objects and space are inseparable. As Rinpoche writes:

> In the mental realm, it is more readily apparent that thoughts and the awareness that makes them available are made up of the same 'stuff'. KTS 150

The reading from KTS explores whether it is legitimate to make this link between the mental and the physical—to see space as encompassing both. Look for ways to explore this question in your own experience. To anticipate a bit, similar questions will come up for time and knowledge: Is there a time more engaged with the heart of 'timing' than linear time? Is there a knowledge that does not depend on models and interpretations?

The main reading for this week is Ch. 3 of TSK. The three views described on pp. 50–51 is a good starting point. Look especially at the new 'definition' of space at TSK 50:

> 'Space' will henceforth refer to a dimension of reality whose openness is both a prerequisite and concomitant feature of all *experienced happening*.

On first reading, this may not be a very easy notion to understand, or even to make sense of. But if you return to it as you do the readings, and especially in light of your experience with the exercises, it may begin to open up. Try writing out the meaning of what is said here in your own words. Does this little exercise bring any greater clarity?

Another challenging statement in Ch. 3 can be found on p. 64, where it is set off from the rest of the text:

Mind, like space, has no foundation.

The sentence that follows is also important: "This is in fact what accounts for the mind's capacity to accommodate experiences." The ability of mind to accommodate experience, like the ability of space to contain objects, could be understood in terms of power or in terms of freedom. Nothing holds space back, and nothing fills it up. If the same can be said of mind, and if we in some sense participate in mind, or express the power of mind, then we too have the ability to enjoy tremendous freedom. The discussion of Great Space has to do with this freedom.

Exercises

The reading assigned for this unit contains a very important statement about the methodology of TSK:

> Exercises 11 through 13 help . . . to develop both a sensitivity to the fact that a certain setting is in effect and also some skill in *changing* the setting. This is accomplished by working with what could be called outputs of the 'mind-setting' *in ways that are uncharacteristic for that setting.* TSK 64

This approach applies not only to this series of exercises, but to all forms of TSK inquiry.

As for the specifics of TSK Ex. 11, there is a very useful statement in the reading:

> Events and thoughts . . . do not actually have the dynamic connections that we commonly assume and attribute to them. TSK 56

This statement contains an invitation. We can stop organizing our experience so carefully, "stop making sense." But how? If we stay on the level of content, the "what's happening" of our experience, we will end up substituting one way of making sense for another, replacing one set of "dynamic connections" with another. TSK Ex. 11 gives us a way to go beneath the content. It reminds us we can focus on the process of thinking without losing ourselves in what is thought. If you have a chance, try going on to TSK Ex. 12, Space Between Thoughts. The two exercises are closely related, and you can explore both this connection and the differences between them.

DTS Ex. 2 is also closely related to TSK Ex. 11. Again, it invites us to go beneath the surface, not busying ourselves with the usual "bookkeeping." The second part of the exercise, which focuses on hearing and seeing, seems to go beyond thinking into the domain of the senses. But ask yourself if that is really the case. How much does thinking play a role in the way we see or the way we hear?

By the way, there is a typographical error in the instructions for this exercise in some editions: Line three should read, "let yourself *rest* in the activity of thinking."

Unit Eight

Alternatives to Linear Time

LOK pp. 109–26, 179–82
Exercises: LOK Ex. 14, 22
Supplementary Exercise: LOK Ex. 13

The chapters from LOK for this unit (14, 15, and 22) mount a direct challenge to linear time. The writing is more straightforward than in the TSK chapter for the previous unit, and it should not be difficult to follow the argument, which is logical and clear. But of course the real point is to understand the practical implications of challenging linear time.

The very fact that this argument is so rationally laid out can mask its power. In some of the readings assigned later in the program, the challenge to linear time is framed as a challenge to the self, and then there is no use denying that our own values and way of life are on the line: If time is different, the self is not what we think it is, and if the self is not what we think, our lives cannot help but change.

Here, however, the argument may seem more abstract. We have grown used to dismissing the value of reasoned analysis as a tool for changing our lives, a point discussed in LOK Ch. 5, already assigned for Unit Two (see also the TSK Guide, Topics 54 and 58). So the real question to ask is how to make this argument into an experiential reality. How would you go about *activating* a sense of time as non-linear?

One approach to exploring this question is simply to become aware of how time passes, and to look for ways to change this typical movement. This is an example of the methodology mentioned in the last unit: "working with what could be called outputs of the 'mind-setting' in ways that are uncharacteristic for that setting." What particular approach you choose is up to you, but try to come up with an approach that you can actually put into practice, keeping notes on what you find in your journal.

Regarding the impact of time, consider this quote:

> If the linear structure of time is open to question, the basis for the world known through descriptive knowledge collapses. LOK 125

Let's apply this claim to our own experience. Is it even possible for the basis of the world—as we experience it—to collapse? Could we live in such a world? Or would we restore that basis the moment it collapses, so quickly we might not even notice any gap? Are the structures we put in place self-sealing, like an automobile tire that cannot be punctured?

These questions are of course rhetorical; the answer is implicit in the close of Ch. 15, where the self (here, "the subject") is identified as the one who imposes order. It is made explicit in Ch. 22. We are order-making, sense-making creatures. That is what we are all about. But to make sense, the world has to unfold in time in a particular way. That is what the challenge to linear time is getting at.

Linear time and the self are intimately related, a relation we will be returning to in later units. This point is made in a surprising way in the reading:

> The threefold division of time is fundamental to direct experience. The being of what exists only in the past is completely different from the being of what exists only in the future, and both in turn are completely different from the being of what exists in the present. In terms of this division, it seems accurate to say that only what exists in the present has 'real' being at all. What happened in the past may once have been real, but it no longer is; what will happen in the future may later be real, but it is not yet.
>
> Something remarkable is at work in the view that the present alone is real, for the 'location' of the present in time is established only by the *presence within it of this particular 'I'*. Thus, the existence of all that is seems to hinge on the *wholly subjective fact* of 'my' being here. LOK 181

The point being made here is in one sense self-evident and trivial, something we never find it necessary to think about. But the TSK vision invites us to look at what we normally ignore. How can it be that reality depends on my presence 'in' it? If reality is bound to *my* subjectivity, how real can it be?

Exercises

LOK Ex. 14 is not that different from TSK Ex. 11, which dealt with space (see Unit Seven). Again, the idea is to go beneath the surface. Of course, to talk of "surfaces" and going beneath is to apply spatial concepts to temporal structures. Is this legitimate? Is the whole idea of linear time another spatial metaphor applied to time? If so, then perhaps the move suggested here is completely appropriate.

Rinpoche presented an exercise closely related to this one in great detail in *Dimensions of Thought* (1980). Here is part of his commentary:

> This way of exemplifying the emergence of more 'knowing' of 'time' is useful, but not entirely faithful

> to the process. For, the noting of smaller and smaller mathematical divisions requires that our individual 'awarenesses' become temporally shorter. Moreover, the quantitative *total* of time that we are aware of would remain fixed, regardless of how many tiny discrete intervals we might register. Neither of these limitations applies to the case of using more 'knowingness' to track 'time'. . . The 'small' interval is not really smaller, nor is the 'larger' one really larger . . . The more extensive duration and unfolding movement of time required to constitute an apparently longer interval can be discovered to be illusory.
>
> DOT 1, 38–42

LOK Ex. 22 refines Ex. 14 by focusing on the point of transition between two moments, an indeterminate 'x'. For a similar exercise presented less formally, see DTS 99; see also TSK Guide, Topics 43, 60. DTS Ex. 13 and 14, both of which will be introduced later in this program, are also relevant.

The supplementary exercise, LOK Ex. 13, raises an interesting question: We remember events largely because of the future that emerged from them. If we strip away that future, do we still have any reason for remembering them? If we have no reason to remember them, what becomes of them. Do they simply disappear from memory?

Unit Nine
A Challenge to Models

LOK pp. 129–44; VOK pp. 91–98
Exercise: DTS Ex. 4
Supplementary Exercises: TSK Ex. 23, 34

This unit completes our second pass through time-space-knowledge with a challenge to conventional ways of knowing. Like the previous unit, it depends on getting clear on what LOK calls descriptive knowledge.

Here is a slightly shortened version of the main discussion of descriptive knowledge in LOK:

> Descriptive knowledge draws on the past to bring essential coherence and order to the direct experience to which polar knowledge gives access. Through reference to established structures, it allows distinction-making and the assigning of meanings. Only through descriptive knowledge can the self know its world as specific objects and events, each with its own distinguishing qualities and characteristics that persist over time.

> Yet descriptive knowledge also places severe restrictions on what can be known. The self understands the content given in each new moment . . . in light of available concepts, prior descriptions, and fixed definitions. Each succeeding definition or description sets in place a 'fixed perspective' with its own 'focal setting', limiting what can be understood. *To define is to be confined within the definition's narrow range.*
>
> Because the self perceives by reference to the past, it is invariably drawn outside experience, which occurs only in the present. For example, if the eye sees the moon reflected in the waters of a river, the 'present' experience is the continuously changing flow of water illuminated by the light of the moon. Descriptive knowing, however, leads the self to identify the content of the experience, naming it 'the reflection of the moon'. That definition is then verified in succeeding moments (aided by a characteristic shortcut that 'makes sense' of each new moment by assimilating it to the previous one except for specifically noted differences). Repeatedly confirming the echo of its initial perception, the self perceives a steady image of the moon, while the 'present' play of light on water recedes from view. LOK 104–106

Reliance on past-centered descriptive knowledge is closely connected to the conventional view that knowledge depends on models and theories. Chapter 16 of LOK explores a traditional skeptical argument against such knowledge: If we can only know what we experience and we can only experience what our concepts and models allow us to perceive, does this knowing or experiencing count as knowledge at all? Notice especially this statement:

> The model *that establishes models as a valid way to transmit new knowledge* is one such model.
> LOK 130

Notice also the comparison of 'facts' in a novel to 'facts' in reality. Here you are, reading these words: Isn't this 'really' happening? In one sense the answer is clearly yes. But the text is pointing to a different level.

The treatment of what models leave out (LOK 132–34) is very powerful. Reflect on it for a while before doing the assigned exercise, which could be considered an antidote to the difficulties described.

The kind of analysis presented here has been used in the Western tradition to fuel relativism and nihilism (compare LOK 139–41). It is important to recognize that here it is being used for a very different purpose. The aim is to undermine conceptual knowledge and thus introduce the possibility that a different kind of knowledge is available.

The challenge in working with this unit is one that operates throughout the TSK vision, but is brought home with special clarity here. How can we follow the invitation to question conceptual knowledge without ourselves getting stuck on a purely conceptual level? Here the material on the witness (LOK 142–44) is especially important. Does the witness succeed in moving to a level that is not conceptual? Be careful not to answer the question too quickly, too conceptually. Let the feeling of the witness sink in *as you are reading*, and then look and see for yourself. You can also invite the feel of witnessing at other times in your experience. What does reality feel like? How do you know? How can you say?

The reading from VOK (especially at 91) suggests a completely different approach to models and interpretations; namely, that they be understood as carriers for the knowledge that they superficially deny. The suggestion is that we look to models, trends, and terms that are new to see what knowledge they hold. Of course, to do this, we have to be clear on how models and interpretations operate, and we need to come to terms with their claims to being "the way things are." All of this is explored in the reading.

One of the major themes in these readings, although it is usually expressed mostly in passing, is that the temporal momentum of conventional knowing, which takes for granted the linear unfolding of time, is an enemy to knowledge. The exercise for this unit explores this link.

Exercises

There is only one assigned exercise for this week. This is because DTS Ex. 4 (Conducting New Knowledge) has six subparts, so there is plenty to work with. One way to think of this exercise is that it starts from the transition point 'x', introduced in LOK Ex. 22 (Unit Eight). The various parts of the exercise proceed in a very structured way to penetrate words, ideas, and concepts to arrive at knowledge. Rather than trying to do all parts of the exercise in one setting, work your way into it slowly. You might find it useful to practice DTS Ex. 4 for a while and then do one of the supplementary exercises, perhaps alternating back and forth between the two. Be sure to allow enough time, since each supplementary exercise should be done for a minimum of 15–20 minutes at a time.

The first supplementary exercise, TSK Ex. 34, has already been assigned as a supplementary exercise for Units Three and Nine. Its significance in the context of this unit is that it completely short-circuits conceptual analysis, and thus serves as a wordless commentary on the possibility of a knowledge that does not depend on concepts and models. If you do the exercise, ask yourself whether it truly does evoke such knowledge. How can you tell?

The second supplementary exercise, TSK Ex. 23 (Going without Going), has also been assigned before, in Unit Two. How does "going without going" relate to "knowing without knowing?" Again, how does this kind of "going" relate to "conducting" knowledge, the practice called for in DTS Ex. 4?

Unit Ten
Field Gravity

KTS pp. 177–89; TSK pp. 6–14, 40–42
Exercises: KTS Ex. 40; TSK Ex. 4–5

With this unit, we start Part Two of the study program, shifting from the 'body' of Time, Space, and Knowledge to the 'experience' of Time, Space, and Knowledge, seen in light of the three levels of the vision.

The TSK reading was also assigned in part in Unit Four, but now we come at it from a different perspective. To understand the difference, consider the following statement:

> If we try to 'track' . . . space *by itself* to any great degree, we may fall into confusion or experience a kind of 'black-out'. While space around objects may seem appropriate, necessary, and restful, space within objects may appear a little bewildering. And space considered by itself may appear meaningless, or perhaps even forbidding.
>
> Space *by itself* is difficult to attend to because our ordinary world view has an implicit preference for tangible 'things'—existence, existents, interactions, causes and effects, and events. TSK 7

Mindful of this difficulty, in Unit Four we inquired into space by looking at the borders and partitions of the objects that appear *in* space. Now we are going to take on space more directly, through the second-level notion of space as a field.

The starting point for this approach can be found at TSK 40–42. The idea that appearance is a "consolidating tendency," emerging from space without ever really 'departing' from space, is central to an understanding of space. In first-level terms, this same point might be put very differently: space contains not only the object being investigated or visualized, but the one who knows or visualizes that object, *in the capacity of knower.*

Of course, this is not our usual understanding of space. But it is useful here once again to recall the definition of first-level space at TSK 111: our ordinary realm, seen in the light of further possibilities. To focus on the contents of space is to lose sight of such further possibilities. To bring in the knower, and to acknowledge that what appears in space depends on the operation of a certain knowing capacity, is to allow for the availability of second-level space. Second-level space is distinguished not by the non-standard experiences it allows (though these are valuable as a clue to its presence and an invitation to deepen knowledge), but by its capacity to encompass a second-level knowing that is never content to rest with a given output or content to experience.

The reading from KTS, our first from the body of this book, introduces and explores the idea of space as field, expanding on what was said in different terms in the readings from TSK. A good starting point is the discussion at KTS 184–85, which presents the notion of a field in the context of familiar, first-level forms of inquiry.

Analysis in terms of the field allows for several new insights: the parallel unfolding of objects and interpretation (178), the availability of lower-level space *within* the field (179), and the role of 'given together' (179). At the other end of the spectrum, the reading presents the failure to recognize the field as field as the source of a breakdown in knowledge (180). This is evidenced by the inability to see the "mechanisms and mechanics" of the field (186).

Still, the very act of pointing out this limitation suggests a strategy for inquiry:

> A knowledge attuned to the 'field dynamic' and not confined by 'field constructs' brings to light the interaction among space, time, and knowledge—understood not just as aspects of the 'field dynamic', but as second-level structuring principles more basic than the 'field'. Alive to this interplay, our understanding of human beings, human concerns, and human interactions could move to a new level of insight. KTS 186

In general, KTS 186–87 helps show how the exercises for this unit support the analysis and bring it alive.

As is typical of KTS, the reading for this unit is dense, with more new insights on a single page than are found in many complete books. If this all seems a bit overwhelming, just read lightly, absorbing what you can. Perhaps reflecting on the many images—for example, the image of 'field gravity'—will offer a way into what is being said.

Exercises

KTS Ex. 40 (Chapter 40) has strong parallels with the two Giant Body exercises (TSK Ex. 4–5) also assigned for this unit, and may help reveal different aspects to those exercises. It starts from the perspective mentioned at the start of the unit (space around objects). In doing the exercise, you will move naturally to focus more on space, making a transition from level one to level two. Both the KTS and the TSK exercises make sense at both levels. If anything, the KTS exercise is more closely bound to level one, while in TSK the transition to a level-two perspective is made an explicit theme.

Unit Eleven
Field Dynamics

TSK pp. 63–68; KTS pp. 190–206
Exercises: Ex. A; KTS Ex. 48
Supplementary Exercises: TSK Ex. 14–15; TSK Ex. 34

As in Unit Ten, the reading for Unit Eleven includes a selection from TSK that has already been assigned. The focus is on the way space 'gives' appearance. A good summary statement is found at TSK 65:

> Great Space does not stand in a two-term relation to its manifestations.

This rather formal vocabulary may conceal the point, which is simply that what appears in space is, from the Great Space 'perspective', not different from space as such. If this is so, it no longer makes sense to speak of space 'containing' its contents, or even of objects, thoughts, and so on 'emerging' from their sustaining space matrix.

It follows that the way space 'gives' appearance is not the way we think. It is not like a magician reaching up and snatching a canary from thin air; rather, the canary itself, like the magician and even like the act of 'snatching from thin air', manifests space. And this is true in the most literal sense:

Appearance makes space manifest without ever becoming anything other than space.

This is not a concept conventional thought can really make sense of. In part, this is because our conventional understanding of space is bound up with our conventional understanding of both time and knowledge. First there is empty space, then something appears (conventional time); we assign an identity to what appears, whereas space is what is without identity (conventional knowledge). Only by challenging all facets of this dynamic interplay do we open the realm of second- and third-level space.

Now, second-level 'space-as-field' is not at all equivalent to pervasive Great Space. But it shares with Great Space the sense of intimacy with all appearance, and this is what sets it apart from first-level space. It might be useful to reconsider in this context the statement discussed in Unit Four: "space is projecting space into space."

The reading from KTS elaborates on this theme of the intimacy between space and its appearances, clarifying what is meant by 'field'. It starts with some very effective and convincing first-level examples of the 'field' in operation (190). The next few pages present a kind of creation myth (though the discussion on 199–200 rejects the idea of creation as too closely bound to conventional understanding). This 'myth' culminates in the claim that "whatever arises 'embodies' or expresses the 'field' as a whole."

In discussing field dynamics, KTS notes that knowing is a part of the field. Compare this to an earlier TSK statement:

> If we employ new 'focal settings' and see the way they work, we can come to *an overall understanding which is itself a kind of space.* TSK 5

Ch. 42–43 of KTS continue the analysis of experience from a 'field' perspective. Given that this whole concept is so foreign to our first-level understanding, they are remarkably clear. Here is a particularly good summary:

> Each 'this' and 'that', each 'here' and 'there' implicates the 'field' as a whole. The object 'at the center' and the directions that radiate out from it are again definable only in terms that involve the rest of what appears. In the physical realm, even the all-important distinction between form and space, through which alone the object can be established, turns out to rest on particular acts of knowing that the 'field' accommodates. KTS 196

Incidentally, this might be a good place to reflect on why the TSK books so often put terms in single quotation marks, sometimes known as 'scare quotes'.

The material starting at KTS 198 is not easy reading. Read lightly, and see what you can do with it, perhaps returning to it more than once. To explore further, take a look at the readings for Topic 67 in the TSK Guide. The basic point relates to the insight that it is our focus on existence that closes off space, and that a field analysis, by focusing on allowing rather than existing, helps counter this fixed orientation.

Even if you do not understand everything you are reading for this unit, keep in mind that no matter how far they depart from ordinary ways of talking and thinking, the TSK materials are not pointing toward some secret understanding or claiming special insight. They are attempts to encourage inquiry and activate a new vision. That is all.

Exercises

Ex. A, first assigned (and described) in Unit One, is a useful way to explore the notion of field dynamics. Here are five possible ways to 'apply' it, though there are certainly others:

- ☆ Expand and condense whatever shows up within the field, possibly looking for qualities that endure no matter how much expansion and condensing take place. Could such properties, if you do find them, become gateways to field dynamics?

☆ Expand and condense whatever is put forward as a candidate for a field mechanism.

☆ Try expanding and condensing the field itself. Is this even possible, or does it require standing 'outside' the field, in another field?

☆ Consider the possibility that through expanding and condensing, a new field is created.

☆ Is there a field within which the very possibility of expanding and condensing arises? Can you expand or condense it?

KTS Ex. 48 introduces the notion of the 'feel' of the field. Because it works primarily with the senses, it can help counter any tendency to move toward abstraction in engaging the significance of "field dynamics and mechanics." Notice the different qualities of knowing associated with each of the senses. For instance, seeing tends naturally to focus awareness and knowing on a specific point: the object *at* which we look. In contrast, hearing quite naturally immerses us more in our environment, especially if we do not focus too much on the meaning of what we hear. The sensation of touch (which includes the inner feeling of the body, known sometimes as proprioception) naturally invites a knowing that is sensitive to our 'situation', in the most literal sense. For more on the senses, see Tarthang Tulku's *Joy of Being* (2005).

There are three supplementary exercises suggested for this unit. It would be useful to take the extra time to do these exercises, because we have reached a point in the program where we are really beginning to explore a new vision of reality, and this would be a good opportunity to consolidate whatever insight or clarity may be emerging. Each of these exercises has been assigned or recommended before, but now you could do them in a way that encourages a field perspective, with a special focus on the observer as a part of the 'given' field. In practicing TSK Ex. 15. Part E is especially relevant to the 'field' perspective.

Since each of these three supplementary exercises deserves concentrated focus, you might try at this point setting aside a day or more for a short retreat. You can structure the retreat any way you like, reviewing the readings, returning to other exercises, writing in your journal, investigating the interplay of time, space, and knowledge within ordinary activities, and so forth. But in the midst of whatever you choose to do, try giving time to each of these three exercises. If weather conditions do not allow you to practice TSK Ex. 15 (which requires blue sky), you can set it aside for another time, or else you could try visualizing blue sky instead.

A note about TSK Ex. 15: To do it in accord with the instructions in the TSK book requires concentrated effort over several days. But it can also be both engaging and nourishing when done in shorter practice periods. The exercise refers back to the statement on TSK 3 that the sky is our local representative of space. Here we are trying to use it is as a gateway to the direct 'experience' of 'space allowing'.

Unit Twelve
Field of Knowledge

TSK pp. 13–18, 91–94, 98–108; KTS pp. 368–83
Exercises: LOK Ex. 48; KTS Ex. 44
Supplementary Exercise: TSK Ex. 34

Having introduced the second-level notion of the 'field', we now explore it in terms of the knowledge dimension. The theme is the relation between concepts and space, and the suggestion is that opening up concepts can be a gateway to different levels of space.

TSK 91–94 introduce a way to carry out such an 'opening up'. The discussion at TSK 98–108 explores in detail the need to activate the vitality of time in order to do so.

The second of the three readings from TSK contains an interesting observation on the relation between models, discussed in an earlier unit, and the TSK vision:

> We have access to [an open and liberating 'space'] dimension as part of our unique heritage as human beings. Nevertheless, our 'attainment' of it depends

> on a departure from usual attainment-orientations, because the usual notions of a 'person' and of 'the world' are too fixed to be suitable for discovering space. By replacing these inflexible views with an alternate model, which may then blossom into a liberating vision, we can realize the joy and fulfillment involved in 'human being'. TSK 91

The passage goes on to explain why, in light of this possibility, it is important to focus on the most fundamental facets of being and to explain the role of 'challenging' in doing so.

Reflecting on what is said in this short excerpt can be very helpful in understanding how to approach the TSK books as an expression of the vision. Some people approach the vision primarily in terms of the exercises, and this can certainly be very fruitful, even liberating. But here Rinpoche points to the need to replace the set of views through which we usually orient ourselves with *an alternate model.* In large part, that is what the TSK books do. They do this not in order to convince us that one model is accurate, but so that we can get involved in a model that challenges the usual views. We can explore it, question it, 'excavate' it until it blossoms into a liberating vision. If we do not do this, we risk interpreting even the most powerful experiences in light of what we already accept as true, and in that case, we set limits on what is possible.

Of course, there is always the danger that we will turn the models that the TSK books present into a new set of convictions and presuppositions To guard against this, Rinpoche delights in presenting an abundance of different TSK models, some of which seem to have little to do with one another. A good example is the model of 'field dynamics and mechanics' that we are working with right now. Do a search on the term 'field' on the CD, and look at how it is used in the first TSK book. You will find some illuminating passages, but you will not find the term used in the same systematic way it is used in KTS. The two books present very different models. In this program, we are making connections between the two models, but this does not mean the two can be reduced to one. Each model reveals a different facet of the vision; each is a doorway into the "liberating dimension" of being the TSK vision evokes.

The reading from TSK 98–108 marks a transition, for we are preparing ourselves for our first in-depth look at temporal facet of being. Time, Space, and Knowledge are interdependent, each implicated in the others, and here the point is made that opening to Space requires activating the vitality of time. The suggestion is also made that the knowledge that comes from pursuing this course has nothing to do with meanings, concepts, etc. The exercises for this unit explore that same point experientially, and it is probably useful to practice them with this perspective in mind.

The readings from KTS focus directly on how the field shapes knowledge and is shaped by knowledge in turn. Chapter 81 lays out the basic points. Take a look especially at the last paragraph, which introduces the idea of 'fields of knowledge'. What does it mean to speak of "the transitional operation" of such fields? In reflecting on this question, it may help to compare the notion of field dynamics and mechanics.

If Ch. 81 of KTS explores the second-level field dynamic, Ch. 83 goes more to the third level of space and knowledge. The change is signaled in this passage:

> Our present limitations, with their characteristic qualities, are based on 'field intensity' and 'feel', and so are not absolute. Ordinary consciousness itself is *determined by the 'field'*. It seems we could cultivate an alternative—an awareness that is free and unknown. KTS 376

Here 'field' should be understood in both its cognitive and spatial senses; indeed, this duality is central to the whole conception of space as 'field'. See in particular the paragraph on 'space' at KTS 377. The discussion of 'field indeterminacy' as *more* than a speculative possibility (379) points directly at third-level space.

Exercises

The short KTS chapter called Fields of Feeling (381–83) could itself be read as an exercise, more general in scope than KTS Ex. 44, but related to it. If you want to try out this approach, see if you can integrate practice of Fields of Feeling with the practice of LOK Ex. 48. In turn, to make sense of Ex. 48, you will have to go back and take a look at LOK Ex. 47. Finally, you might compare TSK Ex. 13 and its commentary, which are not assigned in this program. In doing LOK Ex. 48 focus on the space aspect of being; we will return to the same exercise later with respect to time and knowledge. The instructions for the exercise suggest doing it playfully. By all means take this seriously.

Unit Thirteen
Story of the Self

LOK pp. 169–76, 185–90;
TSK pp. 165–71; DTS pp. 115–25
Exercises: DTS Ex. 11; TSK Ex. 9; TSK Ex. 18
Supplementary Exercises: LOK Ex. 5; DTS Ex. 12

The focus now shifts to time. As the readings from LOK make clear, our usual conception of time is closely linked to placing the self at the center of experience. This is because the self is a story told 'across' linear time. Just as lower-level space and the objects and thoughts it contains are 'given' by a second-level 'space field', so the story of the self is 'timed out' *as a story* by second-level time.

In this model narrative time, shaped by the telling (or living out) of stories, is one possible way for the temporal dynamic to unfold. Notice how different this understanding of lower-level time is from the earlier formulation of lower-level time as an impersonal grid. It is interesting to ask whether and how these two versions of lower-level time are related. Is one more fundamental, more 'accurate', than the other? Do they perhaps serve different purposes?

Chapter 21 of LOK is very clear and penetrating, and makes perfectly good sense read simply in first-level terms.

The emphasis on the stories of the self may seem similar to various psychological accounts of the self, but there is a significant difference. At the risk of over-generalizing, we might say that psychology recognizes the importance of stories, but has a specific strategy for working with them: 1) identify the stories in operation; 2) tell a new set of stories; 3) make these new stories the basis on which the client lives her life. The TSK vision, in contrast, aims to replace the story-centered way of being and living with a vision not centered on the stories of the self at all. The discussion of the 'founding story' in the reading helps make this distinction clear.

Regarding the link between the story of the self and linear time, see especially LOK 173–75. Here is a thought-provoking summary statement of the problems involved (LOK 173):

> The ongoing persistence of a unified self appears to the mind as a self-evident truth. It brings with it two corollaries: the flow of experience from past to future, and the objective existence of a surrounding world that is subject to that flow. The temporal dynamic of the self and of history are parallel, indisputable structures.

The reading from TSK refers back to an earlier discussion of 'timing' (148–53), not assigned in this program. It involves the sense that time 'times out' the content of experience, without being restricted to experience. Here is a paragraph that suggests the practical significance of this 'timing' model:

> Surrendering the ordinary blinding 'knowing' to wider perspectives . . . amounts to feeding everything that is given with 'time' back into 'time'. This can be done quite deliberately—skillfully and vigorously feeding 'time' into 'time', stimulating and accelerating it, getting more resultant energy and again putting that back into 'time'. All understandings and realizations which we have can be fed back in this manner, rather than holding on to them and taking them as the 'truth'. TSK 149

This theme has been anticipated in LOK Ex. 14 and 22, assigned in Unit Eight. It will also be addressed in more detail later in the program. In all other respects, the material from TSK is a perfect complement to the LOK reading. It presents very similar points, but more from a second-level perspective; that is, more from the perspective of time itself.

This unit includes for the first time a reading from the body of DTS. See as you read if you can get a sense for the special flavor of DTS in contrast to the style of TSK, LOK, and KTS, with which you already have some familiarity. The metaphor of conducting, which has already come up in DTS exercises, is central to the main model for temporality in DTS, so take time to reflect on it. Consider the image of conducting time in the same way that a wire conducts electricity. Does this parallel the image of space as field?

Exercises

DTS Ex. 11 is a natural complement to the reading. The reference in the exercise to "second-level" stories may be misleading; it is not meant to evoke second-level time; rather, the "second-level stories" referred to here are more fundamental stories; stories that 'evoke' or 'constitute' or 'emerge out of' (three different formulations that may suggest different avenues of inquiry) the founding story.

TSK Ex. 9 is a space exercise, and it builds on the Giant Body exercises. But it is also an exercise that focuses on the sense of self. The notion of 'coming out', which this exercise introduces, is related both to 'space as field' and to the story-making dynamic that binds us to linear time.

LOK Ex. 5, one of the supplementary exercises (and also a Foundation Exercise) was also a supplementary exercise for Unit Two. If you did it then, go back and look at your journal notes and ask yourself how your experience with it this time differs from your earlier practice. If you did not do it then, see if you can make time for it now, since it will not be assigned again. This exercise makes a very good complement to TSK Ex. 9, and it certainly deepens at an experiential level the analysis in the reading.

If you would like to work with DTS Ex. 12, the second supplementary practice, it would be a good idea to set aside several hours. Notice the tendency described in the exercise to deal with one story by telling another, often insisting that this second story is not a story at all, but simply 'the way things are'. Once you are tuned in to this tendency, you may find it operating in many aspects of your life.

Unit Fourteen
Trading Knowledge for Identity

SDTS pp. 3–14, 17–21; LOK pp. 193–208
Exercises: TSK Ex. 23; TSK Ex. 25
Supplementary Exercise: LOK Ex. 13/18/25

The theme of the self is central in TSK, and each of the books works with it differently. In the reading for this unit, we see how SDTS comes at the self. The SDTS approach, with its intimations of geometry, could be considered more space-centered than time-oriented, but the distinction is not that sharp, for once the axis of the self in relation to its known world has been set up, time radiates out along that axis in the form of a cone. (For some relevant images, see SDTS 120–26.)

If time is a cone radiating out from the present, it no longer makes sense to speak of linear time, since nothing radiates in a single dimension. Compare this to the idea of conducting used in DTS: a wire conducts electricity in a straight line, but a conductor who leads an orchestra conducts in a way that seems to burst the bounds of space. Does the image of 'radiating out in a cone' fit that of the orchestra conductor?

LOK Ch. 24 challenges the self in a systematic way, focusing especially on the witness. The witness as a specific aspect of experience was first introduced in the reading for Unit Nine, where it was linked to a kind of feeling or conviction:

> the 'feeling' of reality, or the conviction that 'this is so'. Wholly apart from subtle philosophical investigation, it is this feeling we rely on in conducting the business of our lives. LOK 143

LOK Ch. 23, assigned for Unit Thirteen, began to explore the witness, which the self claims as its own, and suggested that the self's claim of ownership over the witness is suspect:

> Somehow these various pictures of the self in its different roles . . . seem to leave something out of account. Beneath the patterns of the self—as perceiver, objective entity, narrator, and owner—is the witness. The story the witness tells—the identity of the self—may be 'just another story'. But still there is the witness itself: the powerful assertion that something *is*, with a direct vitality that cannot be denied.
>
> Perhaps the witness is not 'just another story'. Perhaps the force behind its testimony—which operates entirely apart from the content of what it has to say—is a direct expression of the dynamic energy implicit in the flow from past to future and the corresponding vitality of the present. The witness may point to a primary thrust 'within' time that polar and temporal knowledge, together with the manifold structures of the self, do not fully reveal. LOK 189–90

Now that suggestion is explored in more detail, using the phenomenon of the witness as a bridge to make the connection with higher-level time:

> But what is the content to which the witness testifies? Normally we interpret the message of the witness in terms of the identity of self and the substantiality of the world. But perhaps the witness is speaking of something very different. Suppose that beneath the time marked out in ordinary experience there were a 'deeper' time, a fabric into which the patternings presented in conventional time were woven, or else a dynamic whole without distinctions—a time presenting all meanings but prior to all meanings. If the witness were speaking of this deeper time, usual interpretations that we give its message might prove ill-founded and misleading . . . Time, space, and knowledge could perhaps bear witness in a completely new way. LOK 193–94

For the continuation of this analysis, see LOK 197. DTS also has a lot to say about the witness: a search on the CD will let you trace that discussion.

LOK Ch. 25 focuses more on the consequences of putting the self at the center, with a special focus on the founding story. Here is a good summary statement:

> The narrative unfolding of the self's story amounts to an ongoing agreement the self makes with itself for the purpose of witnessing its own identity and affirming its ownership over experience. LOK 207

Exercises

Going without Going, the first exercise for this week, has already been assigned as a supplementary exercise. It allows an in-depth exploration of the claim that if the self is not the actor, time can manifest differently. It would be good to do the practice for 45 minutes or longer at a time. LOK Ex. 25, the supplementary exercise, will be assigned for the next unit, but you may find it helpful to explore it in the context of the themes presented here as well.

Unit Fifteen
Presentations and Exhibitions

KTS pp. 297–317
Exercises: LOK Ex. 13/18/25; TSK Ex. 30
Supplementary Exercise: LOK Ex. 47

The readings and practices for this week are meant to evoke or activate the third level of time, a point discussed quite concretely in KTS 315. In practice, this means putting knowledge, time, and space at the center of our being, instead of the self and its stories. (Compare LOK Ex. 48, assigned for Unit Twelve.)

The relation between the self and conventional knowledge is fundamental. The two arise together, which is to say that the self can be understood as a particular take on knowledge (KTS 298–99). In turn, this means that third-level time is not accessible to the self:

> With [the self's] style of knowing in operation, whatever responds to a rhythm or sequence of time that does not allow for the operation of the self . . . will remain unknowable. KTS 299

Have you had experiences, in the course of this program or otherwise, that might express such 'non-self' rhythms? What might examples of such rhythms and sequences be? Does TSK Ex. 23 (Going without Going), assigned for the previous unit, give any insight into this question?

Looking from the other side, KTS 300 offers a helpful summary of first-level time as seen from a third-level perspective. Can you sense of the power of this alternate way of looking?

The discussion of the 'bystander model' at KTS 307–11 is related to LOK Ch. 24, assigned for Unit Fourteen, but the link to time is made more explicit:

> The 'field' of 'bystander-outsider interaction' possesses a characteristic 'feel' and an intrinsic temporal momentum . . . As soon as an act of observation begins, the 'bystander' projects an inner sense of time. In a single 'moment', the energy of an inner intensive awareness contacts what will be identified as an observed object, situating it through juxtapositions and dichotomies that understand the present in terms of a particular past and future. It seems unlikely that such a momentary knowing will be accurate. KTS 308–309

Consider in this connection the statement at KTS 309–10 that the 'bystander-self' never merges into the "triple field of time, space, and knowledge." Do you have any sense from the inquiry you have been conducting in this program what such a merging might be like? How you might head there? You might say that the supplementary exercise, LOK Ex. 47, puts this question to the test.

The contrast between 'bystander' ways of knowing and a third-level temporality is well expressed at KTS 313:

> Suppose that awareness were not owned by the 'bystander', and so were available 'within' experience, 'as' experience. In that case awareness would

> operate 'within' each mental capacity as well. Available at the outset as well as the outcome, such awareness could be understood as the expression of a deeper knowledge, alive with a vitality prior to distinctions and determinations.

As with so many statements in the TSK books, this brief passage could serve as a practice, available to inquiry at any time.

Exercises

The LOK exercise sequence (13/18/25) is similar in some ways to TSK Ex. 18–21, which you might want to review. There is an interesting discussion of the TSK sequence (including Ex. 22) in the front matter of TSK that has relevance to the LOK exercises as well. Here is an excerpt:

> Through showing us how to 'know' 'time', the practice of these exercises bring many benefits. We can learn to view both positive and negative situations as 'time'; and as such, we can transform them, or even appreciate them as a fulfilling 'sameness'. We can work with 'time' any way we like, and can even directly experience times that would conventionally be designated as being in the past or the future (and therefore not ordinarily accessible). We can develop a better understanding of the locatedness of the knowing self in our limited temporal structure.
>
> We ordinarily relate to life as a stream of transactions and are always 'on the move'. As a result, we are confronted with a realm pervaded by transitoriness, continual shifting and collapse . . . By opening each thought and presentation to Space-Time-Knowledge, the linear cause-effect and transaction view gives way to an appreciation of the vastness of Being within and as each thing. The transitory is then itself not transitory. The moment of death does not involve a real 'passing away'. TSK xiii–xv

LOK Ex. 13 was already assigned as a supplementary exercise in Unit Eight, and it is discussed briefly there; see also Unit Fourteen.

TSK Ex. 30, Subject-Object Reversal, also assigned in Unit Three (and one of the six Foundation Exercises), offers another fundamental challenge to our usual orientation. Since this exercise is being done in a time context, you might wish to emphasize the *activity* of knowing that it expresses. In doing the exercise, how does the object know? Can it know without becoming the subject? Can it know when no story has been engaged to drive the activity of knowing forward?

Unit Sixteen

Positions and Positioning

LOK pp. 223–28; KTS pp. 19–23; LOK pp. 313–16
Exercises: LOK Ex. 41; DTS Ex. 9–10

Over the next three units, our inquiry will center on knowledge, and specifically on the logic of the logos. The logos has already been introduced in Part One of the Program (Unit Six), but now it receives a more extensive treatment.

Just as temporal 'conducting' offered a parallel to the second-level 'space field', so the logos offers a fairly precise analog to second-level space in the domain of knowledge. It provides a 'mechanism' through which one can conceive of a knowledge that is not owned by the self, told by a narrator, or affirmed by a witness.

It may seem surprising that Chapter 27 of LOK, which introduces the logos, does so after a discussion that focuses on confusion and fear. After all, we are talking about knowledge, not emotionality. But that is just the point. The self 'knows' things to be a certain way, consistent with its own identity as self; if they are not that way, or rather, if they could be known to be *differently,* the self will inevitably feel threatened. The TSK vision offers wonderful new possibilities for knowledge and being, but we should not be surprised if the self (and remember: that's *you*!) erects barriers, such as a sense of loss or threat, to realizing such possibilities.

The self's lower-level knowledge is inseparable from emotionality and confusion. It is not just that the self has accepted a particular kind of knowledge, and as a result it is more prone to falling into negative emotionality or to being threatened. That is true enough, but it is a very preliminary assessment. The link is much more fundamental. Emotionality is a part of the structure of knowledge that the self has set into place, a part of "the way things are" that conventional knowledge first sets in place and then discovers and confirms.

This is no small matter. In exploring the TSK vision, we are trying to do something that in our 'selfmost' heart of hearts we do not want to do. If you can, let yourself stew in this dilemma. We want to go beyond self-centered ways of knowing, but first we need to acknowledge that being a self is who we are. For some ways in which this acknowledgement could play itself out, see LOK 225–26. The reading from KTS fits right in with this theme. You can read it lightly for now, because we will be returning to it later in the program.

In reading Ch. 38 of LOK, be aware of any tendency to turn it into a solution to the dilemma explored in Ch. 27. "Positions as positionings" is a key insight, but it is not one that will work at the conceptual level.

Here is a passage from this chapter on the central importance of positions and their role in the structuring of the self:

> The self *establishes itself and the world that it knows* by moving off-center, 'beyond' the point-instant of temporal knowledge. In one unified action *it takes a position, posits a situation, and imposes meaning.* The self finds itself 'in the world'—its attention defines 'there' and 'then' and its vantage

> point locates 'here' and 'now'. Form and feeling, specific characteristics and qualities, knowing and being, and the entire realm of human values and meanings all arise as a direct consequence of the self's positionings. LOK 147

As a way of leading us deeper into the potential for the self to free itself of positions, this chapter relies on the idea of the 'read-out', best understood in this context as another way of talking about the logos. In Unit Six, we looked at the definition of 'logos' in the LOK glossary. The glossary also has a very helpful definition of 'read-out':

> The informing, communicative output by time of a particular focal setting on space, based on a certain sort of knowing being in force. The 'read-out' purports to establish its content as real, but cannot establish itself; nor can it be established by anything 'outside' the 'read-out'. Whatever is put forward as 'outside' the 'read-out' is either bound to the 'read-out' or else takes form within another, independent 'read-out' and thus has neither status nor effect.

Exercises

LOK Ex. 41 works with emotions directly, bringing out their structural, knowledge-bearing component. The practice of switching emotions as a way of dealing with emotionality is deeply valuable in its own right, but the 'structural' element of emotionality, the way in which emotions are given by the logos, is pointing at something different.

The two-exercise sequence from DTS (Ex. 9–10) would be well worth exploring in depth. If you can, try to set aside a period for more extended practice. The progression of these exercises is very important. DTS Ex. 9A is a kind of loosening-up exercise. Focusing on space is a way of turning from the knowing/witnessing activity of the self to a more pervasive knowing that gives access to the logos. The second two parts of Ex. 9 go further toward the sense-making, meaning-giving aspect of our experience, where the logos could be said to operate. DTS Ex. 10, which focuses on what might be called the 'source-and-arising' of the logos, requires extensive practice, but perhaps you can get a sense of what it is about through exploring more lightly. The commentary on Ex. 10 is especially illuminating.

Unit Seventeen
Embodying the Logos

TSK pp. 193–208; KTS pp. 411–425
Exercises: TSK Ex. 26; LOK Ex. 48
Supplementary Exercise: TSK Ex. 25

The reading from the TSK book starts with a strong focus on time, but we are using it to explore the qualities of knowledge. Our specific focus is read-outs, introduced in the previous unit. Again, the reading emphasizes the idea that no one read-out is linked to any other read-out, since this insight makes it possible to challenge more convincingly the linear sequencing of time. Our focus, however is on the knowledge that informs each read-out.

While the TSK chapter is especially concerned to make clear that no read-out establishes anything (199) and that everything can be challenged, our emphasis is a little different. The point that especially interests us right now is that the logos encompasses everything, including whatever justifications might be put forward as proof that something can be established apart from the logos.

As a way into these points, here is the discussion from the reading on the act of seeing and approaching a wall, with its striking claim about what is possible:

> Consider the sight of a wall from across the room. You know that if you were to walk towards it, you would eventually impinge upon it in a way characteristic of walls—namely, it would stop you from moving further.
>
> None of this applies to 'things' seen on the read-out level of the second stage of 'time'. It is true that due to our forward and outward directedness, and our need for continuity, what is presented in one 'read-out' seems closely related to the content of successive 'read-outs'. And certainly the next read-out seems clearly established as such. It makes no ordinary sense to doubt the unbroken temporal connectedness and direction.
>
> But these factors can be doubted on this new level of insight. We do not have to go a certain way. We can travel off the beaten path. This does not mean that 'we' can walk through walls. But knowingness can apprehend an uncharacteristic sequence of time, which would ordinarily be considered as a 'passing through the wall' sequence.
>
> Even when we do follow along an ordinary sequence, the confirming relatedness and regularities are not unchallengeable. The same is not the same. Sameness is a matter of conviction within a read-out. Unbrokenness is not, beyond all question, unbroken. It is just a message that things are or have been a certain way. TSK 197–98

Once we have a better appreciation for the power of the logos, we can turn to our real focus for this unit: the possibility of knowing the logos itself. This is a prospect we explored experientially in the previous unit, using the DTS Ex. 9–10 sequence. KTS calls this special kind of knowing 'cognition': "knowledge *by* the knower *of* the logos." KTS 411

Access to the logos is central in second-level knowledge; at the same time, it opens the possibility of third-level knowledge, which goes beyond 'cognition'. As suggested at KTS 422, one way to make the transition from second level to third level is by focusing on the 'places' "where the old structures break down, where internal inconsistencies and unresolved mysteries come to the fore." The approach being presented here is both subtle and surprisingly accessible:

> Because we can go to such places in our inquiry, we do not need to challenge the temporal order directly. It is enough to 'let' the 'order' speak for itself; we are led naturally to points of transition from 'order' to 'order'—points that serve both as signs of limitations and as symbols that point beyond.

Note here the importance of not knowing, of being willing to entertain mysteries. These mysteries, however, can be very ordinary. For instance, it often happens that we fail to act on our best intentions. Why should this be? Again, we have great difficulty in focusing our minds on one object, or one intention. Is this an example of an "internal inconsistency," similar to those described in the reading?

In this connection, see especially the discussion of the subject-object relationship and its special intimacy in KTS, 422–24. What sense can you make—not just conceptually, but experientially—out of the statement on KTS 423: "The image of the object can be said to understand itself?" Practice of the supplementary exercise (TSK Ex. 25) could be very helpful in this exploration.

Exercises

The suggestion was made at the start of the program to incorporate a body-centered relaxation practice, such as Kum Nye, into your practice. That would be especially helpful in this unit. Alternatively, you could loosen up with one of the exercises we have already done: good choices might be Ex. A, TSK Ex. 24, or TSK Ex. 34.

TSK Ex. 26, Transcendence of Pointings, is described as a follow-up to TSK Ex. 7, Body-Mind-Thought Interplay. Again, the practice is at once very subtle and surprisingly easy. Notice that both the mind and the body can be challenged. In fact, their very appearance or *appearing* (and pointing) is what challenges them.

LOK Ex. 48 (which relies on LOK Ex. 46) could be seen as a follow up to DTS Ex. 9, but it places more emphasis on knowledge. The key sentence in the exercise is this: "With the 'knowingness' of knowledge central, all positionings lose their privileged place."

The supplementary exercise, "Intimacy," is discussed briefly in TSK, 195:

> In Exercise 25, there is a rough approximation of the 'read-out' phenomenon, with the practice of seeing everything that comes up as being 'you'.

Does it make sense to you that this is only a "rough approximation?" If so could you find a way to refine it? Suppose you did this: would this amount to letting TSK Ex. 25 evolve into TSK Ex. 26?

Unit Eighteen
Inquiry as Knowingness

TSK pp. 233–48
Exercises: TSK 29, 32
Supplementary Exercises: LOK Ex. 46; TSK Ex. 34

The term 'knowingness' was introduced in the first unit of the program, and has come up repeatedly since. In this unit, based on Chapter Twelve of TSK, it becomes the focus of inquiry.

The reading opens with a critique of beliefs and concepts, a topic explored in Unit Nine. Here this critique is integrated into a perspective that presents all appearance, all arising, as "all just lower knowledge." TSK 235

A key theme for this unit was already introduced in Unit Seventeen: The knowledge that the TSK vision aims to activate is at once very subtle (even unimaginable), and directly accessible, without special effort:

> We need not take the 'mind', psycho-physical organism, or subject as fundamental. Nor do we need to

> preserve the conventional emphasis on our ordinary knowledge as being a particular functioning or state (and by relaxing this emphasis we can clear away the idea that a better 'knowing' need be a variation—a nonordinary state, such as an 'altered state of consciousness'). By using such a fundamental challenge as a guide in refining our experience, we will find a different type of knowingness to appear of itself . . .
>
> It is not set apart from ordinary knowing by some state-specific boundary; it is relevant because it actually apprehends the same content that we know. The point is that it does so without reinforcing ordinary evaluations and preferences regarding what is given in that content. This different approach to 'knowing' can question the content of ordinary knowledge, and even the status of 'the world', without rejecting anything or claiming that ordinary things do not exist.
>
> TSK 234–35

Read in context, it is clear that the insight expressed here is one that the TSK book has been aiming at for some time (TSK 235: "This formulation finally succeeds . . .") Why has it been so long in coming? What had to prepare the way?

One answer to this question is that we first had to understand the logos and the founding story, and especially the notion of embodying the logos. More generally, the reading for this unit summarizes and pulls together much of what we have been exploring for the last nine units, relating space and time and knowledge in a new way:

> Our realm is lower 'space', 'time', and 'knowledge', with the term 'lower knowledge' being used in a way akin to the particularly comprehensive type of 'meaning' related to 'time' in Parts One and Two [of the TSK book]. In this same sense, we can even say that the world is a 'meaning'. We can then go on to say that the world itself is a 'belief', in a special

> sense related to the unusually comprehensive scope of the term 'lower knowledge'—but not a 'belief' in the ordinary sense that it is just what we have been taught to see or believe in. TSK 236

If you are clear on what is being said in this paragraph, you have a good handle on 'knowingness' on the conceptual level. Of course, conceptual clarity by itself will not avoid the tendency of lower knowledge toward the "tediously repetitive and deadening" (TSK 236). But unless we turn it into a story to hold on to, a possession that affirms our own privileged status, or an excuse for cutting off inquiry, conceptual insight can be a truly valuable guide, showing us where to head. Experiential inquiry can do the rest, helping us stay honest and open in our engagement of being.

The closing pages of the reading, starting at TSK 242, take on the challenge of an experiential inquiry into 'knowingness' and the transition from second-level knowledge to Great Knowledge. Notice especially the discussion of 'mindfulness', defined at TSK 243 as "appreciation of Space and Time rather than a 'minding' fixation on conventional 'things'. Those of you familiar with the term 'mindfulness' in the context of Buddhist practice will probably find this a rich and challenging observation.

Exercises

TSK Ex. 29 works directly with the notion of a continuum of knowing, a concept introduced at TSK 240. TSK Ex. 32 offers no conceptual bridges to the discussion of 'knowingness' in the readings, but it embodies knowingness nonetheless. If you have the opportunity to schedule a more extended period of practice, this would be a good time to do so: perhaps a day or a weekend or longer. You could work more intensively with these two powerful practices, as well as the supplementary exercises for this unit and TSK Ex. 23. At some point you might also want to extend TSK Ex. 32 into TSK Ex. 33, Self-Transcending Appearance.

Even though our focus in this unit is knowledge rather than time or space, do not feel too strongly tied to this division, which remains mostly conceptual. In particular, if you work with LOK Ex. 46, try 'putting' Time or Space at the center of experience and see what happens. You might try doing this exercise for a fairly extended period, taking several breaks to write in your journal. As for TSK Ex. 34, the clarity that this exercise evokes is inseparable from knowingness.

Unit Nineteen

The Dynamic of Time at the Center

KTS pp. 3–30

Exercises: SDTS pp. 82–87 (Investigating Time); SDTS pp. 214–15 (Alternative Geometries)

We have now reached the mid-point of the program. In the remaining eighteen units, we will investigate time, space, and knowledge more intensively, devoting six units to each. In the next six units, we will read the entire section on time from KTS, including some material that has been covered in previous units. In the following six units, devoted to space, we will read the entire section on space from DTS, supplemented by readings from the other books. In the final six units, which investigate knowledge, readings are drawn from five of the six TSK books.

The opening three chapters of KTS, assigned for this unit, present the three times, a topic we have explored in Units Five and Thirteen. You might want to review your notes from those units. The basic approach is one familiar from past units: a description of our ordinary experience, presented in such a way that it opens up other possibilities and alternatives. Experientially, the aim at the outset is to 'contact' time directly (without prejudging the issue, raised in earlier readings, of whether or in what sense this is even possible). The end of Ch. 3 marks the transition:

> If 'time' is not specific moments of time, it must be an entity separate from those moments. Can we make sense of such a possibility? In doing so, what assumptions about the 'nature' of time will we apply? Can those assumptions be further investigated?

The first exercise from SDTS, described below, provides an experiential basis for making this transition.

The three chapters assigned from KTS are collectively called "Experience of Time." Note that "experience" is given a precise definition at KTS 20: the reality known to the self. Normally we use the term more loosely (for instance, in speaking of experiential inquiry), but this more precise meaning makes clear that 'experience' is inherently limited, since it is tied to the central structure of the self. Ch. 4 of KTS begins to explore that limitation, using familiar terms such as 'bystander', 'logos', and 'read-out'. Look carefully at the discussion of how experience intensifies time, and how intensification leads to identification. The basic idea here is that 'experience' leads us off in the wrong direction:

> Since the 'order' imposed through (the self's) experience is not centered in time's dynamic, it is only superficial. A sea of emotions seethes beneath the surface, occasionally erupting. KTS 20

This link between lower-level, experience-based time and emotionality is well worth exploring in depth.

The two chapters from KTS that follow continue to explore the self's relation to time, focusing on the relation between subjective and objective time and the structure of the former. Here is a key quote:

> The self can appropriate the dynamic of time as the substance of its own being. KTS 26

In reflecting on this statement, it may be helpful to compare a striking sentence from an earlier book:

> The force of the self's desire unfolds as the momentum of linear time. LOK 149

Exercises

Investigating Time, the exercise from SDTS (presented on 82–87) has four parts:

A. Feel of the Image (82–84). The practice is to explore the characteristic 'feel' of a memory, asking how the memory gets identified and labeled.

B. Drawing Closer to Time (84–85). The practice is to inquire into the difference between the feel of specific past memories and the feel of time's 'pastness'. This is a subtle distinction, but keep in mind that asking the question is not meant to yield the 'right' answer, but to evoke other possibilities of experience.

C. Investigating the Future (85–86). This part of the exercise repeats Part A, with the focus on the future instead of the past.

D. Edges of Time (86–87). With the focus still on the future, the emphasis is on contacting the dynamic of time: 'now' as it moves into the future. Compare DTS 99, which reverses the direction of flow: "To rediscover the immediate feel of [the connection to the dynamic of time], we can go directly to the point of arising itself: the point in each experience where the future could be said to come into being."

The second exercise for this unit, Alternative Geometries, comes from SDTS pp. 214–15 . The term "ratio," used in this passage, is one we have not met with before. For our purposes, we might think of it as the specification of the logos; for more, see the TSK Guide, Topic 23. In any case, the focus of the exercise is relatively narrow: it consists of the first full paragraph on 215:

> Rather than remembering or forgetting the past, we can inhabit it. Inhabiting it, we can let go of it. Letting go of it, we are able to specify the future, not at a distance, but from within. The tenseness of our present way of being can give way.

Inhabiting the past does not mean hallucinating that we are now in a past situation, but it does mean not approaching the past from outside. Just be willing to wrestle with how this is possible. In terms of SDTS, the ratio provides one possible approach, but it is not one that we can go into in this program. Perhaps later you will be inspired to investigate this special *Sacred Dimensions* approach on your own.

As supplementary exercises for this week, you might want to return to exercises from previous units related to the structure of the self. Keep in mind, however, that the focus of our inquiry at the moment is on time, and on what it means to experience time, rather than on the self. Edges of Time, the last part of the first exercise for this unit, offers a very straightforward approach to this question.

Unit Twenty

Rhythms of Allowing

KTS pp. 31–61
Exercises: DTS Ex. 13; TSK Ex. 24
Supplementary Exercises: TSK Ex. 23; LOK Ex. 46

The reading starts with a provocative statement: "As an aspect of experience, linear time could be said to arise as the projection of a particular awareness." KTS 31

This statement is related to the statement from LOK quoted in the previous unit: "The force of the self's desire unfolds as the momentum of linear time." Remember that the term 'experience' is being used here in a precise, almost technical way. Experience in this sense is not different from our ordinary understanding of experience, but (to paraphrase material discussed in earlier units), it is experience seen in the light of other possibilities.

There is an interesting question here. Suppose you think this statement through in light of the reading and your own inquiry so far in this program, and decide that it is 'true'. How far does this insight get you, if you continue to 'experience' yourself as subject to the firm grip of linear time: the endless passage of moments that make up past, present, and future? The conceptual insight itself does not really change things. But if you could actually 'experience' time as the outcome of a particular awareness, or of desire—or even get a sense of how this may be so—then things may change rather dramatically. Of course, for this to happen, 'experience' may have to change as well.

Another surprising statement occurs a few pages on in the reading: "We are therefore never truly *active* in time" (KTS 33). Be careful not to agree with this claim too quickly. On the surface, it seems outrageous. Occasionally it may indeed seem to us that we are being swept along by events, but this is rare. Most of the time we *feel* quite strongly that we are acting in time. And the reading itself tells us that mind is active, and that the bystander-self is active. Is the point that we are cut off from direct experience? Or does 'experience' itself take us outside of time?

The discussion that follows the statement just quoted suggests ways to counter the condition of 'not-being-active', ways in which human beings can interact with the dynamic of time or be active in time. The next chapter continues this theme, looking at the usual modes of experience and temporality in terms of the logos. The suggestion being made is that when we tune in to the *dynamic* of time, the structures projected by the self—the temporal order that makes experience meaningful—lose their solidity. This does not mean that they somehow becomes transparent in any ordinary sense. Instead, they are no longer convincing in the same way.

The second half of the reading for this unit consists of the section of KTS called "Time of Knowing." The interplay of temporality and knowledge explored here invites a radical shift in our way of being. What might it be like to engage a timing not bound to concepts? The discussion of language in KTS Ch. 10 is important for this question, since it points toward precisely this alternative.

The element of rhythm introduced in the readings for this unit will be explored further in the next unit. Here is a passage to reflect on:

> [Second-level] rhythms of becoming allow for fluctuation between something and nothing, inviting a knowing that goes deeper than polarity and contradiction." KTS 59

What is meant by a rhythm of becoming? Are the rhythms at work in time like the rhythms in a piece of music, or is something else being invoked?

Exercises

DTS Ex. 13A offers an immediate inquiry into the dynamic of time, and also introduces the experience of temporal rhythm. You might try doing the exercise once or twice before you study the commentary; then, after reflecting on the commentary, do it again.

TSK Ex. 24, Marriage of Sound and Breath, is one of the Foundation Exercises. Through its focus on the breath, it evokes the sense of fluctuating rhythms suggested in the reading. When we tune in to the rhythm of breathing, the rhythm of experience spontaneously changes, and because our usual patterns of thinking and sensing—even the contents of our thoughts—are closely linked to this rhythm, experience also changes in a fundamental way. Bringing in the qualities of sound encourages this shift, since a focus on sound tends to de-center the bystander-self that experiences.

Both the supplementary exercises have been assigned before. Each invites going beneath ordinary self-centered structures to contact time's dynamic directly.

Unit Twenty-One

Rhythm and Measurement

KTS pp. 62–76

Exercises: TSK Ex. 24, 25, 26

The reading for this unit relies on the concept of the logos to clarify the distinction between first-level 'order' and second-level momentum. A striking image related to this point is used at the start of the reading, based on the image of a robe woven by time and designed by knowledge:

> At [the first] level, the fabric itself—its material, properties, and origin—could be said to remain unknown. The effect will be as though the design and pattern of the robe were preserved only in photographs of a lost original: The 'structure' or 'order' of both fabric or pattern are still 'available', but there is a loss of a deeper, aesthetic dimension, in which the texture and weight of the fabric and the intricacy of the weave are also available to be appreciated.
>
> This restriction disappears as the momentum of second-level time becomes more directly available.

> The arising of knowledge is seen as given by this momentum, as if the fabric were a tapestry, its design emerging in the course of the weaving. KTS 63

Reflect on this contrast in your own life, not just by looking back at past experience, but through a careful inquiry into the operation of your own mind and the flavor and quality of your own experience. A practice like TSK Ex. 7 might be a good way to ground this inquiry. Is it fair to say that your life consists of happenings in which a dimension of richness and depth has been lost or remains unavailable? Do you find in your experience the twofold deception described at KTS 63?

The questions raised by the reading go to the deepest level. Can we really conclude that first-level realities are built up out of second-level 'shadows' or echoes (KTS 66)? What is the evidence for agreeing or disagreeing? The best evidence would seem to be special experiences that seem more rich and satisfying than ordinary experience. But a similar distinction may be available even at the very ordinary level. For instance, when you get angry, do you experience the full force of the anger, or do you channel it into agitated thoughts? Is this the kind of distinction being referred to here?

For now, you may just want to hold these questions open. It may take some time before an answer emerges in which you have true confidence.

The previous unit introduced the phenomenon of second-level temporal rhythms. Now the rhythms of the temporal dynamic are explored further. The suggestion is made that rhythm can be seen as responsible for appearance:

> The object becomes the specification of the rhythm, as the first-level temporal order is the specification of the dynamic momentum of second-level time. KTS 68

You are not necessarily being asked to see 'reality' in these second-level terms, though it is possible that there could be some unusual perceptions as you do the exercises. More important is the recognition that first-level orders and objects can be questioned when seen from second-level perspectives.

At this point the parallel between second-level space and second-level time should be quite evident. For both time and space, first-level structures are shaped by the insistence on the reality of what is pointed out. That insistence distorts a more dynamic and allowing vision.

The treatment of gravity and 'indensification' (KTS 73; compare KTS 20) is a useful way to make this point. Look carefully at the progression from gravity and indensification to measuring out and 'pointing out'. For more on points and pointing, see the TSK Guide, Topic 42.

Exercises

All three exercises for this unit have been introduced before. They form a natural sequence, but notice that Ex. 26, Transcendence of Pointings, is placed in a different chapter, whose title refers to the TSK vision as "self-challenging." Does this mean that 'experience', in the technical sense we have explored in earlier units, must be challenged? Or does the challenging go even deeper than that?

In doing the three-exercise sequence for this unit, try making a special effort to integrate it with the reading. How can you make these exercises into an occasion for inquiry into the second-level dynamic and momentum of time?

Unit Twenty-Two
The Motion of Time

KTS pp. 77–87
Exercises: DTS Ex. 14; LOK Ex. 15
Supplementary Exercises: DTS Ex. 16; LOK Ex. 26

The reading for this unit focuses more closely on the energy and dynamic of time once it has been liberated from the 'indensified' structures that center on the self. Although the language being used is quite unusual, by now you have enough background so that the reading may strike you as surprisingly clear and direct. If not, do not struggle too hard; just work with what you can understand at any given time, and return to the text more than once.

A good summary statement of the theme for this unit can be found toward the start of the reading:

> The character of existence is built up out of and through the momentum of 'time'. KTS 78

Compare this to the formulation at KTS 84: "What 'is' emerges as the echo of itself." It probably does not need saying, but this is not meant to be an interesting philosophical insight, but a lived experience.

It may seem self-evident by now that time is in some sense dynamic, even though there is a strong tendency in our usual way of thinking to take time as something static, a series of measurements that divides reality into days, hours, seconds, and so on. Still, there are many ways to approach this dynamic. The momentum of linear time, through which the self emerges, is one form of that dynamic, often associated with a sense of time as bearing down on us until it becomes almost unbearable. There can be a clear sense of pain associated with this momentum, as discussed below in connection with the exercises assigned for this unit.

However, we can also relate to the dynamic of time—even time experienced as linear momentum—in terms of an always available energy:

> As the product of a 'specified' momentum, existence 'has' that momentum as its nature. The momentum that existence 'embodies'—whether as things or as qualities and meanings—is still available within the 'order' of what is. All that is necessary is to thaw the frozen structures that the first level takes as given.
>
> KTS 79

The reading suggests that each of the three times (past, present, and future) can be a gateway to this energy. As you reflect on how this might be possible (with special reference to the suggestive comments at KTS 81–82), focus especially on the past, which in ordinary terms has slipped away into the realm of 'was', lifeless and inert.

Exercises

DTS Ex. 14 explores the dynamic of time directly. In part B of this exercise there is a reference to the 'pain' that comes from being 'stuck' in linear temporality. If you want to explore this dimension of linear temporality, look at the discussion at

DTS 295–98. The following passage, at the start of that section, may be enough to get you started:

> When we lose contact with time, we have cut the dynamic central to our lives. Like a spinning top that wobbles as it slows down, we live erratically. Subjectively, there is the sense that time is flickering, like a film not properly adjusted on its reel or a flag flapping in a stiff breeze. There is a strain that goes nowhere. Events do not synchronize with one another—the timing is 'off'. KTS 295

If this description of time as "flickering" does not seem true to your experience, it may be because of the progression described just beyond the passage quoted: "Pain [is] transformed into the fixed and rigid structures of linear time." Conversely, this suggests that opening up to time's momentum may mean touching a kind of pain that we normally do not notice. Is this true to your experience?

LOK Ex. 15 engages the temporal dynamic by challenging the usual first-level structure of 'past-giving-rise-to-present' through which linear time unfolds. It would be a good idea to extend this exercise by adding in LOK Ex. 26, listed as one of the supplementary exercises.

DTS Ex. 16, the other supplementary exercise, continues to explore the phenomenon of pain. Feel free to take some time on your own to explore this approach. It can be used to deal with physical or psychological pain at any level.

Unit Twenty-Three
Echoes of Zero

KTS pp. 88–104

Exercises: TSK Ex. 27; TSK Ex. 28

Supplementary Exercises: LOK Ex. 32; LOK Ex. 33

In this unit, we make the leap into third-level time. The concepts in the readings are challenging, but coming at this point in the program should be fairly clear. Just let the presentation do its work of opening up established patterns. Of course, a certain degree of not knowing, even if it initially takes the form of confusion, may be welcome.

The challenge to existence mounted in these chapters has already been anticipated in the reading for the previous unit. Now a new vocabulary is introduced, centered on zero. As you read, you can almost feel the language straining to point where language cannot go, to gesture beyond conventional ways of speaking and thinking.

The concept of the 'zero point' is introduced as something toward which second-level understanding moves, without ever arriving there. But the suggestion is made that without relying on understanding, a different kind of experience may be possible:

> In approaching this 'zero point', we can sense a quality in appearance, a hidden 'occupancy' of time that ordinarily merges in the instant occupancy of experience. This quality could be said to be 'prior' to the structure that identifies 'moments' or designates 'outsiders'. It might be understood as a gateway, a 'first' moment of presence, of space that is pure and undisturbed. KTS 89

Compare this passage from SDTS:

> To open the point of knowledge, we would have to recognize it as the zero-point. At the very beginning of conventional, first-level knowledge, the zero-point is available—the prior of perception. To return to that point is to engage the openness of space, to accommodate experience differently. SDTS 149

Incidentally, a search on the CD for the term 'prior' might prove helpful in thinking through these issues.

Here is another sentence from the start of the reading. It is very dense, but has already been anticipated in the previous unit's discussion of existence "as its own echo":

> We might speak of invariable time presenting the 'singularity-point manifestation' of the conditional existence of what will appear as its own image. KTS 89

The discussion that follows helps bring this statement down to earth. The "merging of Space and Time in intimacy" is something we can look for within (or 'prior' to, or beneath) experience. It can take us to the heart of experience, bringing to life the dynamic adventure that is 'happening' in the text.

As you read further into the text, do not struggle too hard to understand each idea. The language is evocative, and you might approach it in the spirit of DTS Ex. 2B: "understanding what is being said without relying on the words." Remember that TSK inquiry aims to stay light and playful, an adventure of discovery open to everyone who engages the TSK vision.

We have been investigating rhythm in the last few units, and the reading for this unit introduces a wholly new understanding of rhythm:

> We might say that time's rhythm is built up externally out of the basic momentum through regularities of motion, but intrinsically is more accurately the flow of a luminous energy independent of motion . . . : dynamic without the need for motion. KTS 92

Notice how strongly this description, like the rest of the reading for this week, relies on the possibility of a new way of experience, on tracing for ourselves "the rhythm through which form evolves out of emptiness" (KTS 92). We have seen repeatedly that TSK reading can itself be an exercise, and that reading can also suggest exercises to be explored at other times. That is particularly true in the reading for this unit.

Exercises

TSK Ex. 27 and 28, assigned for the first time, make use of very ordinary elements of experience, but put them in an experiential context that allows for cracking open their superficial familiarity to arrive at something very different. The experiences that arise in doing these exercises do not necessarily bear the mark of insight, but they do have a distinctive quality linked to the dynamic of time. As you work with them, a better understanding of 'zero-emerging rhythm' may manifest, even if it cannot be directly named.

The supplementary exercises build on exercises assigned for the previous unit. LOK Ex. 33 in particular evokes more directly the 'rhythm without movement' that the reading seeks to evoke.

Unit Twenty-Four
Invariable Time

KTS pp. 105–116
Exercises: DTS Ex. 19, 20
Supplementary Exercise: TSK Ex. 23

In the past several units, we have focused strongly on the dynamic aliveness of time. The reading for this unit, building on the idea of motionless rhythm introduced in the previous unit, counters any tendency to conceive of this dynamic as 'heading somewhere'; in other words, as involving a subtle version of the first-level 'from/to'. This point is made in various ways in the reading. Here is an example:

> Existence, with its steady temporal flow of continuous becoming, may manifest a 'rhythm' of time that *in its depths has nothing to do with momentum.* Changes in shape and form, as well as in underlying energies, may simply express an atemporal 'mechanism' through which an invariable time 'presents'.
>
> KTS 106–107

Perhaps it will help clarify this point to go a little more deeply into an example touched on in the text. It is a common, everyday occurrence to 'arrive' at an event that we have marked on the calendar some time earlier. For instance, you make a dinner date for next week, and eventually the time for the dinner arrives. But how does this actually happen? True, we experience time passing, but the time that passes, the moment to moment arising of experience, has no intrinsic connection to the calendar time through which the anticipated event arrives. It is just that as moments give way to more moments, clock time is also ticking away in a predetermined sequence. Does it make sense to think of calendar time, viewed in this light, as an atemporal sequence?

This way of looking at time as invariable has immediate consequences for our usual model of 'arriving' at insight or clarity (for instance, insight into the 'meaning' of the TSK vision) through a process. Rinpoche does not reject the value of a process of inquiry. Instead, he suggests that we approach that process differently:

> We can invite Great Time by remaining on the path of inquiry that investigates conventional time . . . We can discover Great Time by allowing it to arise in the fullness of understanding the temporal order.
>
> KTS 113–14

The text goes on to suggest that as we conduct this inquiry, we may come in contact with a 'unitary' time more fundamental than past, present, and future. If you have the opportunity, you might want to compare the discussion of 'nuclear time' in DTS. Here is an excerpt:

> Nuclear time is not one-pointed in the conventional sense. It is not originating time or founding time, though this is how it may appear from within the conducted order. Nuclear time moves, and in its movement includes past and present and future, death and life; includes us all. Without participating in the identities of what is included or has been conducted, it does not negate, nor does it negate negating. Operating before each measured-out move, prior to directionality, it is the not-not-not. DTS 149

This is the last unit on time as it is presented in the TSK vision, so it offers a good opportunity to take stock on a practical level. We have been dealing with a rich set of concepts and ideas, many of which challenge at a very deep level what we long ago learned to take for granted. But of course, the exploration takes place not only at the level of theory. KTS 111 has an interesting passage on this point, one that goes to the issue of 'TSK methodology', first explored in Unit One:

> If we ask how the 'required' change in consciousness is to come about, we can remind ourselves that the 'subjective' mind of ordinary consciousness is not simply theoretical or philosophical or analytical. The subject inescapably acts, interacting with its world and embodying this interaction. For consciousness to change, our embodiment must change.
>
> This does not mean that a new reality must be found, so that consciousness will have something to embody 'into'. Instead, consciousness can be transformed in embodying fully the knowledge that is already available. Seeing that each successive stage of understanding is only another presentation by Great Time is an allowing of this embodying. Time presents in accord with circumstances, not bound either to momentum or to 'order', always inviting Knowledge anew. As Great Knowledge embodies in consciousness, Great Time makes its appearance felt.

Is this description consistent with your experience? As you explore the TSK vision, how do you react? Do you find time and space differently available in your life? Do you have places that feel stuck? Subtle transformations? If confusion comes up, how do you deal with it?

Exercises

The two exercises for this unit are quite different, but each in its own way challenges the momentum and mechanics of the temporal order: onc by 'redoing' that momentum, and the other by inviting a different outcome to the temporal process, as well as a different way of processing. DTS Ex. 19 could be practiced in second-level terms, based on the insight that no one temporal order is fixed. But it is also possible to do the practice in third-level terms. In that case, the momentum manifested by events as they unfolded in the past can itself be 'opened'. This is a different kind of opening—a different 'instant replay'—than one that focuses on the content of the stories we tell and inhabit.

Unit Twenty-Five

A Challenge to Substance

TSK pp. 71–88; DTS pp. 3–8

Exercises: TSK Ex. 7; TSK Ex. 8

For the next six units we focus on space. We do so by reading the entire section on space in DTS, activating a momentum of inquiry in a way that we have not done in looking at space before. We are following along with a systematic presentation of Space as a facet of Being, and as we engage this presentation, many of our taken-for-granted assumptions will—as always in the TSK vision—come up for investigation.

This unit starts with a reading of TSK Ch. 4. You might think of it as background reading for the DTS-centered inquiry that follows. The opening of the chapter (71) announces that the stakes in this kind of inquiry are high:

> When we begin to challenge the 'body', we must be prepared to abandon all sorts of half-measures, and must include 'the world order', the fundament and touchstone of reality, in our deliberations.

The rest of the chapter could be said to make good on this promise, starting with an extended discussion (71–73) of the 'bystander' (as well as the 'outside-stander', referred to in KTS as the 'outsider'). We have already been introduced to bystanders and outside-standers in early units. Given their central role in this reading, it may be useful to compare the following 'definition' from the LOK glossary:

> 'BYSTANDER' The position adopted by the self as an expression of the lack of unity between subject and object. On a first level, the 'bystander' does not participate in the world that it observes and owns. On a second level, the 'bystander' does participate, receiving and 'enjoying' experiences. Within this mode of participation, the dichotomy of subject and object persists, reflected in the interlocking polarities of 'empty/not-empty' and 'exist/not-exist'. For the 'bystander' on the second level, there is a complete separation from space and time and knowledge: Time has no beginning, middle, or end, space displays the endless reduplication of the same, and knowledge can be known only in ways that conform to the presupposed. This separation finds expression in the structures of process, 'from/to', and limitation.

TSK 80–82 and 86–88 lead up to a 'given together' approach already introduced in regard to the 'space as field' perspective.

The reading from DTS starts from the very beginning, giving us a chance to take a fresh look at space, without regard to all the subtle and challenging approaches we explored in earlier units. Of course, the investigation we have been doing lets us look at this material in more depth.

Two key ideas are advanced in this reading. The first is the suggestion that "substance is a mysteriously condensed form of what in our ordinary way of speaking we would call 'nothing at all'" (DTS 5). The second, which complements it, is that any existing 'thing' depends on the boundaries that separate it from everything else. Although this discussion is framed in terms of physical objects, the same point applies to other kinds of entities, such as imaginary objects, thoughts, etc. In fact, it can be very interesting to reflect that when you imagine something (for instance, a car you used to own), you are also imagining the space that that object occupies. What is the relation between the space that you imagine and the space 'within' which the imagining takes place? Or does this question confuse two things of an entirely different nature? Does this way of approaching the matter contribute in any interesting way to the suggestion made at the end of the DTS reading that there might be a more fundamental 'nothing' than the nothing of space-emptiness?

As you read this material, keep an open mind. Although we have explored the 'nature', mechanics, and dynamics of space repeatedly, we are starting a new inquiry, and we do not know where it will lead. Perhaps everything we thought we understood will be called into question, and we will have to start over. This not-knowing is always the right starting point for TSK, for if we think we know the TSK 'take' on experience, we have already lost the power of the TSK vision. As Rinpoche writes (KTS 282), to say "I know" means "I stop here."

Exercises

The exercises for this unit can be approached from this same "don't know" perspective. TSK Ex. 7, Body-Mind-Thought Interplay, is an old friend, but it has a built-in capacity for freshness. The reading assigns this exercise special weight:

> For the sake of attaining an integrated understanding of the new vision being presented, we need to extend the trend of our analyses in Exercises 11 through 14 to the body-mind-thought interplay discussed in Exercise 7. TSK 71

TSK Ex. 8, The Translucent Person, is a direct experiential exploration of the question raised in the DTS reading about the nature of boundaries.

Unit Twenty-Six
A Challenge to Identity and 'Isness'

DTS pp. 9–28, 267–76
Exercises: DTS Ex. 7; TSK Ex. 28
Supplementary Exercises: KTS Ex. 40

The two related chapters from DTS assigned for this unit develop a careful, sustained argument, key elements in which should already be familiar. These include the suggestion that we can investigate space by looking at substance, and that when we do, we find that substance proves surprisingly space-like.

In considering 'the whole of experience', the reading introduces the term "field communiqué":

> The individual appearances that make up the content of experience manifest *in terms of* the field communiqué, which unites them in somewhat the way that stock ownership unites the shareholders of a corporation. Emerging from the field, entities appear hand in hand. They enact a special intimacy that allows for characterizations, distinctions, and an interplay

> expressed through temporal 'from-to' structures. In terms of the field communiqué, the realness of the real is simply the proclaimed authenticity of this mutually co-referring identification DTS 17

Notice how the communiqué is linked to the use of names and concepts, which in turn are linked to the self as knower. (DTS 19–20). A natural further question is how the concept of the field communiqué relates to the concept of the logos.

An important methodological issue for TSK inquiry is presented a little further on in the reading:

> One possibility for cutting through the limitations of a verbal way of knowing is to turn to experience directly, inviting its immediacy to open our senses to a deeper knowing. Methods are available for activating a different kind of awareness, in which we note qualities, characteristics, and interactions without assigning them back to objects that possess them. For example, we might identify a red color, a specific shape and texture, and a pleasing fragrance without making the normal connection to the name 'rose'. At a more subtle level, we could even allow this naming tendency to operate, without accepting its conclusion as final. DTS 21

While this approach has the advantage of offering a systematic methodology anyone is free to cultivate, few people will be able to master it without extensive effort. Again the question arises: Can we only invoke new knowledge through such exceptional means? Or can we discover knowledge in experience as we know it now?

The reading clearly suggests that there is a different way to proceed, one that does not require the same "extensive effort." But before feeling smug because in this program we are involved in this 'superior' kind of inquiry, we might ask whether the structure of the program is subject to the same challenge. Have we been trying to go to direct experience? Or is our focus something different? Just what are we trying to accomplish, and what is our methodology for doing so? These are not rhetorical questions. Your sense of what we are about is bound to affect your practice of the assigned exercises.

The DTS alternative to the effortful and burdensome inquiry we have just been looking at is presented (in one formulation) on 26–28. It is described as playful, and it is based in part on the idea that any limits we encounter in our inquiry and our way of making sense of the world are themselves expressions of knowledge. In a poetic turn of phrase, the text says (27) that this form of inquiry offers "the other side of birth and the beneath of zero."

Finally, consider this cryptic summary statement:

> Without departing from zero, without requiring a body, the exhibition can serve the purpose. DTS 28

In light of all our previous inquiry in this program, where similar themes have emerged quite often, what is your take on this claim? Or do you need to have a 'take'? Perhaps the point instead is that no positions are necessary. In that case, what is required is not so much a specific understanding as an approach, a style of inquiry, and a focus. Space is here presented as offering that objectless focus.

Exercises

DTS Ex. 7 is closely linked to the issues raised in the readings. Notice that the exercise is specifically related at one point to the practice of expanding and condensing, which we have called Ex. A, and that at another point it invokes the Giant Body exercises of the TSK book. Even though Ex. 7B relates to time rather than space, it would probably be useful to bring it into your practice as well. What are the differences? What are the similarities? Looking at the questions raised above, is the point of Ex. 7 to go to the level of direct experience? If not, what *is* the point?

TSK Ex. 28 was also assigned for Unit Twenty-Three. As you do it this time, keep in mind the discussion in the text about proceeding 'playfully'. One way to do this might be to do this exercise simultaneously with another. Feel free to come up with your own approaches, not straining to be different, but not afraid to strike out on your own.

The supplementary exercise, KTS Ex. 40 (at 188), offers a variation on the Giant Body. It also continues the focus on partitions and borders introduced in the previous unit.

Unit Twenty-Seven

Space as Active Allowing

DTS pp. 29–41; KTS pp. 150–54

KTS Ex. 28; DTS Ex. 4

Supplementary Exercises: TSK Ex. 15; KTS ex. 32

The reading for this unit starts with the surprising suggestion that we have allowed space to disappear:

> Our usual focus on substance and identity has turned space into the nothing of non-appearance. Setting space in contrast to what has form, turning it into the simple emptiness of the vacuum, we have made it into a 'formless body': a 'thing' that has the quality of being nothing at all. Misconceiving its nature, we have made it disappear. What is more, since what is without form is without properties, we have assured that space cannot be known. Thus we have made it disappear a second time. To negate this twofold disappearance, we seek now to restore space to appearance.

> Of course, it would be worse than pointless to attempt this restoration by bringing space from the sphere of the unknown into that of the known. If we assigned an identity to space, we would remove from space its capacity to accommodate appearance, and in that case we would no longer be dealing with space at all. If space is to appear, it cannot do so as an object for knowledge, in the manner of things that appear in space. A different knowledge is called for: one more basic than the distinction between known and not-known. DTS 30

It seems fair to say that this program aims in part at allowing space to reappear as a focus for concern. Has it had this effect for you? How do you know? What evidence can you offer? Can you identify specific differences in you actions? In your outlook or perceptions?

As a way of activating the recovery of space, the reading suggests looking at the relationships and the limits of what appears 'in' space:

> We encounter 'space unknownness' in the relationships that space allows, from 'big and small' and 'in and out' to the interplay of subject and object and body and mind. It may even be available in the mystery of our limitations—the constrictions and confinements that we accept as intrinsic to the ways we live. DTS 32–33

The suggestion is made to consider these appearances as "the emerging echoes of the field communiqué." Such a view undermines substance, and this turning away from substance in turn allows space to appear:

> No longer serving as storehouse for what has been labeled and pronounced into existence, space carries appearance into abundance. Ready to exhibit without limitation, it restores room for existence without restriction. DTS 33

The meaning of "eknosis," introduced at DTS 39, may seem unclear. Here are two key discussions, from DTS 39 and 41:

> For space to project into mind, mind must project a knowing capacity out into space. If we cultivate a clear awareness, this outward projection of knowledge can remain available 'inwardly'. Let us call this 'inward outward-knowing' "eknosis." . . . Knowing inwardly and outwardly in a single movement, eknosis moves without departing from its space origin. It allows us to know with a knowledge that is not ours—a knowledge that expands beyond restrictions and condenses into universal availability.

Note that the term "eknosis" may be related to the term "ekkenosis"in Christian theology. The theological term is essentially a variant on "kenosis," and refers to the "making himself empty" whereby Christ abandoned certain aspects of his divinity to become the servant of humanity (and thereby, according to one commentator, made himself subject to time and space). By extension, it has been said to involve the emptying of the ego in reading poetry. Such associations may enrich the significance of eknosis in TSK, but it is probably a mistake to insist on the connection to the theological term, because in TSK the focus is strongly on the knowing capacity that comes with eknosis.

The reading from KTS supplements the DTS reading. It comes at the themes for this unit from a different perspective, and so may help clarify what is being explored. Don't try too hard to reconcile the two approaches; instead, let them illuminate one another.

Exercises

Before doing DTS Ex. 4, read KTS Ex. 32 (Silent Partner, KTS 155–57). It will help clarify the sense in which practice of the DTS exercise can be regarded as an inquiry into space. In the exercise, focus on the aspects you find most interesting now; you may also want to review your journal notes from the last time you did this practice. If you want to experiment with actual practice of KTS Ex. 32, you could approach it as an extension or deepening of the assigned exercise.

KTS Ex. 28 (134–36) invites a series of reflections on space. To take it deeper, consider the closing sentence: "With practice, several 'layers' of space and objects can be envisioned simultaneously in their interaction or lack of interaction."

The other supplementary exercise, TSK Ex. 15, has been introduced in earlier units. It would be a wonderful practice to do if you can set aside time for a weekend or day-long retreat.

Unit Twenty-Eight
The Zeroless

DTS pp. 42–50; KTS pp. 241–52

Exercises: KTS Ex. 56; TSK Ex. 34

Supplementary Exercises: TSK Ex. 25, 26

Chapter 6 of DTS starts with the notion of flattening space. Some people have a negative reaction to this, because we usually associate a 'flattened' version of something with less richness and aliveness: like pinning a butterfly in a specimen case rather than letting it fly free, or like the map instead of the territory. But that is not the idea here at all.

With no properties, dimensions, or measurements, what is left in flattened space? Though we can hardly say that there are things or substances, we should not assume that this leaves 'nothing at all'. To conclude in favor of nothing imports into our 'flattened' world a notion based on the conventional sense of space as opposed to substance, and this misses the significance of flattening. Recall that since we inhabit space, the act of flattening would affect us along with everything else that appears in space. Assuming we made it through this flattening operation 'intact', we might notice no difference in what appeared whatsoever.

Still, the nature of flattened appearance would be transformed at a fundamental level. Freed from dimensionality through flattening, appearance would also be freed from the persistent claims of substance. It would become zeroless.

Flattened space is space in which the claims of dimensionality, which in turn form a particular order and structure that limits and confines, have been punctured. And as the passage just quoted suggests, flattening also affects the dimensionality of the self.

The rest of the DTS chapter moves from the image of flattening to the image of the zeroless. The two images are linked by the central role of the self:

> Measuring and judging, [the self] presents the perspective of its perceptions, proclaiming substance and solidity. Embarking on subjective determinations, it affirms limitations that depart from the wholeness of the whole. Claiming and proclaiming, it takes up and maintains. As these operations expand their scope and enlarge their claims, dimensionality radiates outward—a claim of the self, establishing the closed world of solid identities that a flattened space would never allow. DTS 47

It might be useful to compare the implicit 'from-to' structure identified here with other discussions of 'from-to' (see topic 31 of the TSK Guide, or do a search on both 'from-to' and 'from/to' on the CD). At the same time, since we are talking about projections 'from' the mind, you could compare the discussion of eknosis, assigned for the previous unit.

The material at DTS 49–50 offers a stimulating treatment of how physical and mental space are interrelated. Compare this discussion to the more concrete image used in KTS (which, as a bonus, offers a very different take on 'flattening'):

Subject and object could be thought of as two structures that come together at a single point, giving rise to the known world—like two pyramids that touch one another at their shared apex. If we start from this interaction at the apex, there are no distinctions—only the knowledge of 'two' and 'interaction'. Subject might be object and object subject, for no directions have been established.

Suppose that each pyramid were thought of as a projection of space. From a first-level perspective this would be like flattening the pyramid out. What was in the foreground merges into the background, and the point at the apex completely disappears. Subject and object have vanished, but so has any possibility for knowledge. There is 'nothing at all'.

From a second-level perspective we might think of the pyramids 'dissolving' into space. Though the pyramidal structure itself would remain, there would be a new focus: the way in which 'space' gives subject and object together, leading to their 'interpenetration' or 'co-determination'.

From a 'third-level' perspective, open to Great Space, seeing the pyramids as 'space projections' entails no transition at all. Nothing 'comes from' space and so nothing 'retums into' space. At this level, we must give up the idea of subject and object as the outcome of a causal sequence. KTS 243

Exercises

KTS Ex. 56 (Co-relation, 253–55) focuses on the relation of subject and object, and is closely integrated with the themes for this unit. TSK Ex. 34, which we have worked with before, invites an alternative to subject/object knowledge. Does the experience to which the Embodiment of Knowledge exercise gives rise 'fill' space? Does it transform space?

The supplementary exercises, Intimacy and Transcendence of Pointings relate most closely to KTS Ch. 54 (245–48). For example, here is the last paragraph of that chapter:

> It would be a fundamental mistake to think that Great Space can project only emptiness, for emptiness as the absence of 'something real' is a wholly inadequate symbol for Great Space. Such limitations on 'where' within appearance Space can 'appear' (for example, in physical space, but not in tangible objects, or in tangible objects, but not in thoughts or interpretations) only denies what is spacious in Space.

Unit Twenty-Nine
Symbolic Reality

TSK pp. 6–10; SDTS pp. 3–11, 32–37, 48–60

Exercises: DTS Ex. 5; KTS Ex. 36;
SDTS pp. 223–24 (Balance Point of Knowledge)

Supplementary Exercises: TSK Ex. 34;
KTS Ex. 36 (pp. 173–74)

In this unit we depart from the pattern of working our way through the Space Section of DTS. Naturally this involves taking a different perspective on space. However, the readings for the last unit, with their focus on zero and the point, provide a natural bridge to the readings for this unit.

The brief reading from TSK, already assigned twice, should be read with a focus on what it says about symbols (See also TSK Guide, Topic 21). Here is the key statement:

> Conventional dichotomies themselves preserve, in a symbolic or 'inspired' sense, expressions of a higher order unity. TSK 9

This emphasis on a symbolic, inspired level in appearance also sheds new light on the intriguing statement made at TSK 3: "The sky is a familiar—we might even say 'local'—representative of space."

The first reading from SDTS has also been assigned before. In light of the TSK reading, can "I am here" be read as a symbol of the time-space-knowledge inquiry? What about '16'? Is symbolic reality the same as zeroless reality? The same as 'flattened' reality? Recall the phrase quoted from KTS in Unit Twenty-Two: "What 'is' emerges as the echo of itself."

As you reflect on the symbolic aspect of experience, you may want to consider anew what it means to introduce, engage, and activate a new *vision* of reality. Does a vision operate at the level of symbol? Put differently, do symbols take on their meaning in the context of a vision? DTS Ex. 5 can help clarify this point, for if appearance is zeroless or symbolic, the distinction between a thought and 'reality', which this exercise explores, becomes much more elusive. Abiding in thought may not be different from abiding in reality.

The other SDTS readings continue the theme of exploring the nature of reality. You may want to include in your reading SDTS 29–31, which refer back to the Giant Body exercises in the TSK book. While there are terms in this reading (as well as the assigned reading) that may not makes sense without reading earlier parts of *Sacred Dimensions*, you can let those slide by.

The issue of vision is raised expressly at SDTS 33–34, where "symbol" and "vision" are treated as almost equivalent:

> Of course, the image of a zero-point unfolding in sixteen directions cannot readily serve as a model for the way that substance appears. Instead, we might call it a vision. As a vision, it offers a way of seeing that stands in contrast to our usual understanding of what is so.
>
> Why would we want to employ such a vision? Because our ordinary way of seeing leaves out of account all that is most vital in appearance . . . But the inner truth of 'I am here' can become the gateway to deep knowledge and genuine openness.

This visionary, symbolic reality is here identified as a superficial expression of a deeper order: "the magical body of space." If this 'space body' is also the "inner truth of 'I am here'," engaging the realm of the symbolic and visionary should let us draw closer to the truth of our own being. The exercises can help us explore for ourselves whether this is so.

In the final reading from SDTS, 'zero' seems to serve as a symbol. Notice the suggestion (SDTS 48) that there may be a third kind of space, neither occupied nor unoccupied. Would it make sense to consider this the realm of the symbolic?

Exercises

DTS Ex. 5 has already been discussed briefly above. The exercise from SDTS is freely adapted from SDTS 223–24. We can call it Balance Point of Knowledge. Here is a rephrasing:

> Visualize a circle of intense white light. Let that circle be the wholeness of space, expanding to encompass all appearance. Let the radiance it manifests be knowledge, and let the dynamic of its manifesting be time. When the circle has expanded so far that it vanishes into fullness, visualize it again, and repeat the cycle. As you become familiar with this rhythm, imagine that the whole of being is in balance at that all-encompassing zeroless circle, which is also the 'point of choice' given by knowledge.

This exercise may remind you of LOK Ex. 46 or TSK Ex. 34, a supplementary exercise for this unit. You will probably want to read and reread SDTS pp. 222–24 before you do it.

As for KTS Ex. 36, Catching Mind, the second part is most relevant to the focus of this unit. The first part of the exercise is more closely related to themes that have been explored earlier, and also to issues involving the interplay of space and thought that will come up in the next unit.

Unit Thirty
Dimensionless Thought

DTS pp. 51–70
Exercises: DTS Ex. 6; LOK Ex. 47

The preceding two units have gone very deeply into the visionary aspect of the TSK vision. The readings for this unit in a sense come back 'out', drawing us into our more ordinary way of perceiving and making sense of the world. The intention is that these ordinary structures can now be penetrated, or rather, can be infused with vision.

The reading starts with the point that the thinking mind has structured a world in which space has disappeared. Thus, to recover space requires 'rethinking' thinking. Here the key point has to do with the parallel established in the text: as substance is to space, so content is to thinking (DTS 53):

> Thoughts structure experience by 'building up' reality. Together with their content, they communicate the substantiality of that content—a sense of persistent qualities and independent 'presence': a special kind of 'mass' that exerts its own gravitational pull.

> This gravitational force in turn shapes what appears 'to' and 'through' the mind in ways that structure the whole, conforming to the established order.

We are always thinking, but if we could discover the activity of thinking as sourceless and zeroless, we could return *within* thought to the TSK vision. Initially this is presented as impossible on 'structural' grounds:

> In presenting their content, thoughts refer their own identity to this content. In the act of establishing, they withdraw. DTS 54

Western philosophy sometimes refers to this as the transparency problem: If I investigate what it is like to see a blue patch, all I find is the blue patch: the perceptual act seems not to be available apart from the content of what is perceived.

Although the text identifies other difficulties in inquiring into thought (DTS 55–56), it offers an alternative, linked to the discussion of symbol and vision in Unit Twenty-Nine:

> Thought presents its content without establishing what is presented. Nothing arises, yet this non-arising appears. There is no need for this to be so, but this is how it is. And since it is so, it is vitally important to 'point out' that it is not so at all.
>
> This way of talking is not a trick meant to establish a particular point or point of view. There is nothing to be established, for to establish we would have to proceed point by point, and then we would miss the point . . .
>
> Then how can we respond more appropriately? By activating appearance as symbolic: appearance as zero and zero as zeroless. In this modality, thoughts become the symbols of their own arising. DTS 57

The truly elegant discussion in Chapter Seven of DTS is 'retold' at a deeper level in Chapter Eight. The text looks beyond thoughts to the more encompassing level of narrative. The result is a very precise presentation of just how it is that ordinary reality arises, a dynamic summarized at DTS 61–62:

> Substance takes form through stories that translate zeroless space into dimensionality . . . Identity moves to the center of what appears. Space becomes zero and disappears; mind becomes zero and withdraws. Knowledge also becomes zero, vanishing into the domain of subjectivity. Conventional space and time emerge, together with objects that appear in space and time and the subject that engages those objects. And not-knowing prevails.

Look carefully at the discussion of interactability and multiple dimensionality, and at the claim that these are more than 'subjective' aspects of experience.

The discussion of inquiry at DTS 65–66 offers an opportunity to explore this underlying theme at a deeper and more refined level. For starters, look closely at the first sentence of the section that begins at DTS 65:

> As inquiry moves from speculation and analysis through imagination and embodiment, its emerging dynamic retraces the steps that found the ordinary.

Can you trace the process of inquiry through each of these stages: from speculation to analysis to imagination and finally to embodiment (still considered as an element in inquiry)? Or is this taking too literal an approach? To help you explore this question, you might do a search on the terms 'speculation' and 'imagination' on the CD. These terms figure especially in VOK, which we have not yet dealt with in any depth; we will look at them in Unit Thirty-Three.

Exercises

DTS Ex. 6, a follow-up to one of the exercises from the last unit, starts with the stream of thoughts, then moves on the structures we use to give meaning to experience, and finally invites you to go beyond those structures through a focus on the experience of the senses. It would be a good idea to make each of these three phases central during different practice sessions. You might also find it useful to go back and forth between practice of the exercise and the assigned reading, letting each inform the other.

LOK Ex. 47 was a supplementary exercise for Unit Fifteen. Note the invitation it extends to investigate at the symbolic level, a link back to the theme of the previous unit. In practicing this exercise, can you bring into play the four 'stages' of inquiry identified at DTS 65 and discussed above?

Unit Thirty-One
The Witness

LOK pp. 193–98; DTS pp. 102–114
Exercises: LOK Ex. 16, 22
Supplementary Exercise: LOK Ex. 23

The final six units in the Study Program deal with knowledge, starting with an analysis of conventional structures of knowing. You should have enough background and experience to be able to analyze the dynamics and mechanics of these structures with considerable skill and insight, using the assigned exercises to penetrate the focal settings these structures generate. The last three units emerge once more into the third-level fullness of the TSK vision.

The readings for Unit Thirty-One take time as their focus, but we are approaching them in terms of knowledge. The interplay between these two facets of the vision can serve as one of the unspoken themes for the inquiry.

The chapter from LOK, assigned previously for Unit Fourteen, relies strongly on the concept of the witness, so you may want to go back and review both Units Nine and Fourteen; see also the TSK Guide, Topic 61. Here we are interested in the possibility of a 'deeper' vision of reality, presented in the LOK chapter in terms of time:

> Normally we interpret the message of the witness in terms of the identity of self and the substantiality of the world. But perhaps the witness is speaking of something very different. Suppose that beneath the time marked out in ordinary experience there were a 'deeper' time, a fabric into which the patternings presented in conventional time were woven, or else a dynamic whole without distinctions—a time presenting all meanings but prior to all meanings. KTS 193

At the close of the chapter, this same suggestion is presented in terms of knowledge:

> The founding story recognizes an actor, events, and the unfolding of the narrative. The self unites these elements, possessing experience and organizing it into meaningful patterns. And the founding story asserts that this owner in turn is identical with the witness, the one who verifies the identity of the self.
>
> Perhaps we have misunderstood the knowledge to which the witness testifies. For the witness can only be witness if it is more than another story told within the fundamental structure of the founding story. When the self appropriates the witness, sweeping it within the story-telling structure, it may be bearing false witness. In that case, the true source of the creative energy that the self seems to manifest might be found elsewhere, in a realm yet to be explored.

The question for us at this point is the availability of this "unexplored realm." Reflecting back on previous units, we can ask: "Does the witness bear witness to the zeroless?" The exercises can help make this question an experiential inquiry.

DTS Ch. 12 also deals with the witness, questioning its testimony in terms that we have explored in previous units:

> It is true that time as we know it excludes knowledge from appearance at a fundamental level. But shifting

> to other temporalities may not require a complete revamping of consciousness at all. DTS 111

Again, the question is whether the discussion of the unfounded and the zeroless in which we have been engaging allows for this kind of shift:

> Why feel compelled to 'go' anywhere at all? Within the present immediate arising, we are free to turn inward. In the immediacy of presence, a single thought, a single connection, a single unexpected phrase or turn of inquiry may reveal a new knowledge. DTS 112

As the text notes, "turning inward" does not mean turning toward subjectivity:

> Turning inward has nothing to do with subjectivity. In the inwardness of time there may be nothing to witness, but this does not mean that the witness and what is witnessed collapse into chaos. Instead, it is the substantiality of the witness that gives way. 'Witness' becomes a summary term for a sequence of temporal events, conducted into being in such a way that a particular form of knowledge arises.

Compare the discussion of eknosis in Unit Twenty-Seven.

Exercises

LOK Ex. 16 could be understood as a practice in multiple dimensionality, a theme presented in the previous unit. Here the idea is to look 'beneath' the content of experience to its intensity and aliveness, and to be able to discover these qualities within ordinary perception. LOK Ex. 22 and its follow-up, Ex. 23, offer ways to go beneath the stories that the witness claims to authenticate to something more fundamental and intrinsically knowledgeable.

Unit Thirty-Two
Not-Knowing

VOK pp. 83–90, 157–63;
LOK pp. 379–92; TSK 148–58
Exercises: TSK Ex. 29; TSK Ex. 31
Supplementary Exercises: TSK Ex. 30, 32

In the preceding unit, we looked at the possibility that the witness bears false witness; that its operations actually point toward a fundamental knowingness not linked to the self. Now we turn to not-knowing with a similar question, asking how it can be a carrier for a deeper knowledge. Not-knowing is generally presented in the TSK vision as the opposite of closing down knowledge prematurely. In this way, it serves as a fundamental gateway to the vision, and to each level of space, time, and knowledge. As VOK 158 puts it: "In the search for knowledge, ignorance is not the enemy."

The first selection from VOK explores the prospect of a transition from knowledge owned by a self to knowledge manifesting within time and space directly (compare LOK, Ex. 46, which we have worked with before). The second selection focuses more directly on not-knowing (159):

> Knowledge founded in not-knowing is no longer an object mysteriously emerging in time: a creation out

> of nothing whose genesis remains obscure and whose credentials are thus always in doubt. When we accept not-knowing as source, we can investigate and understand without compulsion, for knowing is no longer an unnatural act. We can proceed more loosely, even therapeutically, letting knowledge emerge naturally.

The use of the term "therapeutically" here may come as something of a surprise. To explore this idea further, you could do a search on the CD for the stem "therap."

The LOK reading suggests that the conventional distinction between knowing and not-knowing does not hold up, or at least that it is not where we want to focus. Instead, not-knowing can be seen as a freedom from knowing what is laid out in advance, from knowing the limitations on knowledge:

> Knowing our not-knowing, we can explore, analyze, and understand without establishing, asserting, or maintaining. Unconcerned by the limits on knowing that our inquiry discloses, knowledge itself can continue to deepen and expand, acting as its own witness. With no fixed starting point, no impenetrable barriers, and no ultimate, knowledge will be unfettered. Things are left 'the way they are', but they are not that way. The witness witnesses no 'ness'. LOK 383–84

In making sense of this claim, take some time to think through the relation between 'conventional' not-knowing and the more profound not-knowing being discussed here. If the two are not related, are we just involved in word games? The last statement—The witness witnesses no 'ness'—is highly relevant, but since we are not going into the concept of 'ness' in this program, you will have to pursue this on your own.

More generally, the reading from LOK raises many important issues about how knowledge is possible and what limits on it seem to arise almost spontaneously. Methodological issues also return to the foreground. Compare the discussion of whether the TSK vision offers a "path" (LOK 387–88):

> We can frame an inquiry into time and space by looking at the assumption that what is being presented here is a 'path' to greater knowledge. In terms of linear temporality, this may be an accurate statement, but phrasing it in this way establishes a structure that puts new knowledge 'somewhere else'. As we have seen, once such a dichotomizing structure has been established, the gap between antagonistic poles cannot be bridged. Distance is determinative: What is situated 'somewhere else' is simply out of reach.

The LOK selection closes with the suggestion that allowing 'room' for alternatives to conventional knowing, space, and time may help us discover "new dimensions" of being. Is this just loose talk, or is it an accurate statement? The DTS descriptions of multiple dimensionality and the zeroless, and the SDTS vision of a sacred geometry are both relevant in assessing this question.

The reading from TSK is again framed in terms of time, but can be recast in knowledge terms that echo those found in LOK: Within ordinary appearance, when that appearance is 'un-known', new possibilities (dimensions) are available.

Exercises

TSK Ex. 29, Awareness as a Reflective Surface, was assigned once before, in Unit Eighteen. It challenges the self-as-knower model directly, and invites you to discover knowledge or knowing where ordinarily you would not expect to find it. TSK Ex. 30, A Subject-Object Reversal, supplements this approach. If you want to go into the same idea from a different perspective, see LOK Ex. 5.

TSK Ex. 31, Not-Knowing as Knowing, offers a natural progression into not-knowing itself. At some point in practicing it, you might want to start by first practicing TSK Ex. 32, A Unifying Clarity. Experimenting in this way with a clarity of vision may help penetrate the usual, automatic equation of not-knowing with darkness.

Unit Thirty-Three
Tracing Knowledgeability

VOK pp. 6–13; LOK pp. 269–310
Exercises: TSK Ex. 7–9; DTS Ex. 4B, 4G
Supplementary Exercise: LOK Ex. 34

The previous unit's readings suggested the possibility of finding knowledge within the unknown. In this unit, we ask more directly how to go about doing that. In a sense, we are returning again to the theme of TSK methodology, which we looked at to start the program and have been returning to from time to time ever since.

If the excerpt from VOK, which focuses on imagination, speculation and questioning, had been assigned earlier in the program, it might have seemed to operate strictly in first-level terms. Now, however, you may notice a deeper connection to knowingness. For instance, consider the connection made in the reading between TSK inquiry and light (a theme also explored in Unit Thirty-Five):

> Like the sun shining in space, [knowledge] is available everywhere, illuminated through its own luminosity.

> We can bask in this light without feeling obligated to trace its arising or discover its source. The questions presented here could be seen as an exhibition of [this] light of knowledge. From this perspective, they lead nowhere beyond their own self-presentation. VOK 11

The extensive reading from LOK starts with a critique of reason as a vehicle for knowledge, based on the fact that reason is a tool for the bystander. Instead, it suggests that we explore directly the way the bystander knows:

> Reason can be investigated in ways that do not depend on reason, and the operation of cause and effect can be opened to inquiry without placing that inquiry within a causal sequence. We can look at the patterns of intentional knowledge without first accepting the confines of conventional intentionality, and can investigate desire without being caught up in its flow. Through such inquiry, new questions can be asked in a new way. In the domain that inquiry opens, does reason still apply? Do cause and effect still operate? We do not know the answers. In this not-knowing, a possibility for a new knowledge emerges. The terms we use and the thoughts we entertain now have a new and wider context; instead of being givens, they are constructs to be investigated further. LOK 270

Consider especially this challenging statement:

> It is not a question of learning new techniques or practices, but of giving up unquestioning loyalty to the techniques and practices that we are *already practicing* in every moment. LOK 274

Ch. 34–35 of LOK go on to consider some of the obstacles and barriers to activating this kind of inquiry.

The penetrating and insightful quality of inquiry introduced in the first part of the LOK reading is balanced by a very different quality that comes into play later in the reading (LOK 308):

> In the activity of questioning, the balance between 'subject' and 'object' calls for bringing into play a certain sharpness and cutting quality. But if inquiry is to call into question the 'question,' the 'questioner,' and the 'act of questioning' itself, it must be more than simply sharp and cutting. For such an inquiry, the question must 'contain' within it the potential for transforming the relationship between 'subject' and 'object'. The move toward separating and penetrating must be 'balanced' by a move toward embracing and embodying. *The fullness of inquiry will open the structure of knowledge when its intention incorporates an appreciation that can deepen into intimacy.*

The section in the LOK readings on Self-Reflection (300–301) is a very important statement of a fundamental aspect of TSK inquiry. Pay special attention to the use of the word "trace" or "tracing." To get the full sense of what is meant here, you might want to do a search on the CD for these terms, either in LOK or in all the TSK books. You could also look at how the term 'tracing' is used in the TSK exercises.

Exercises

Consistent with the focus of the reading for this unit, the sequence of exercises from TSK (Ex. 7–9) is meant to provide a concrete example of inquiry in action. Do the exercises 'facing' in two directions at once: looking at the specific focus of the inquiry, but also focusing *on* inquiry. This is a fairly sophisticated operation, but at this stage in the program it should at least make sense.

DTS Ex. 4B calls for this same kind of bidirectional focus, while Ex. 4G invites such meta-inquiry (inquiry into inquiry) directly. You can of course work with the other parts of Ex. 4, but for the purposes of this unit you may find it more helpful to focus just on thoughts.

LOK Ex. 34, one of the supplementary exercises, has not been assigned before. Now that we are approaching the end of the program, you might think of it as a good review. What does it mean to speak of "more" time or space or knowledge? What does it mean to "invite" them? Can your not knowing the answers to these questions support a different way of being?

Unit Thirty-Four
Knowledge Knowingness

TSK pp. 159–62, 165–71, 263–71, 281–90;
KTS pp. 440–51, 496–99
Exercises: VOK pp. 133–34 (Trinity of Inquiry);
DTS Ex. 23B, 24
Supplementary Exercises: TSK Ex. 23B, 34

The first two readings from TSK (one of which was assigned for Unit Thirteen) chart the link between intimacy and knowingness, and then suggest how knowingness is 'toned down' in our ordinary way of being.

Notice the emphasis in the TSK readings on our own present way of being as the gateway to Great Time and Great Knowledge (159–60):

> Intimacy *is*. It is the shattering—and yet natural—surfacing of our real *Being*. It applies to us; it is not a perfect intimacy elsewhere, beyond the turmoil and isolation of the ordinary world. It is 'here', the only difference being that a Great Space and Great Time view is taken rather than a 'self' and 'thing' view.

The second two readings from TSK will be assigned again (in part) in the next unit. In one sense, they are relatively straightforward, but they also are asking us to make a profound shift in our way of being, a shift toward "love, intimacy, and fulfillment" (TSK 263) that was anticipated in the previous unit.

In reading the discussion of love in the TSK book, it is important to be clear that we are not talking about love at what might be called a 'psychological' level. This is because 'love' as it is invoked here ultimately does not involve the question of whether 'I' love 'other beings' at all. Instead, love is directly related to knowledge: For love to operate fully, knowingness must be available. And knowledge at this level does not belong to or emanate from a self.

On the other hand, there is plenty of material in these readings that is deeply valuable at a more psychological level. In particular, look closely at the discussion of the three layers involved in love—and all other feelings—and reflect on the suggestion that "different mixtures of such strata comprise what we consider our personalities" (TSK 266).

Those of you who have studied Buddhism may have heard it said that love, in contrast to anger, etc. is not an emotion. Here a different approach is presented. Love at the first level, even though it is positive, has a quality shared by all emotions, including those we would usually consider negative. Note that this first level, the 'surface' layer, is the only one the self can perceive. It would be very useful to explore this threefold structure in your own experience, and see whether it is consistent with what you discover.

Although it may seem very different, the following passage from the KTS reading (447–48) could be understood as a commentary on love in the TSK perspective:

> A lower-level indication that knowledge is freely available is the 'co-existence' through which existence points beyond itself to a more comprehensive

whole. The 'coexistence' of existence with nonexistence gives each entity as embodying its opposite; the 'coexistence' of existence with space gives openness in all that appears; the 'coexistence' of existence and knowledge gives the interplay of intention, action, and consequences that brings life to being.

Exercises

The exercise from VOK (Trinity of Inquiry) is freely adapted from the first two paragraphs of the section called "Letting Go." It might be phrased in this way:

> Let the flow of thoughts and experience arise naturally, looking for the sense, the implicit claim, that what you are experiencing or thinking about 'exists'. Whether you experience this directly or not, after a while begin to relate to what arises instead in terms more natural to time, space, and knowledge. As experience emerges, merge with its emerging, As space appears, greet it in boundless wonderment. As stories are told, read out the knowledge their telling proclaims. Relax all efforts to make progress, to possess or control, to succeed, or to understand. Enjoy the unbounded energy of time and the silent clarity of inquiry.

You might try moving from this exercise directly into TSK Ex. 23B, one of the supplementary exercises. After a while, you could try doing both exercises simultaneously.

DTS Ex. 23B returns to the theme of the unknown, but turns it toward the arriving of knowingness. The suggestion in the text for DTS Ex. 23A may be helpful: "Do not expect to turn the unknown toward knowledge; do not expect feedback on your experience that you can make sense of."

Unit Thirty-Five

Light of Knowledge

TSK pp. 281–86; DTS pp. 199–223
Exercises: DTS Ex. 21–22;
DTS pp. 212–13 (Appearance Only)
Supplementary Exercise: TSK Ex. 32

We turn in this unit to a model that investigates the link between knowledge and light. This material needs to be appreciated and investigated in its own terms, but it is important at the same time to keep sight of the underlying theme of pervasive knowingness.

The reading from TSK is a review of material assigned for the previous unit. It offers a bridge between knowledge and light in terms of 'clarity':

> We have seen that open-ended clarity can be found even within the obstructing presence of ordinary knowing—and in ordinary 'not-knowing' and confusion, as well. Both 'ordinary knowing' and its correlate 'not-knowing' can be thawed or cultivated to yield this luminescent quality. TSK 281

How literally should we take this image of clarity, with its link to luminescence? Usually we think of clarity in the sense of physical light as valuable because it makes something stand out with more definition. But to say that clarity itself 'carries' knowledge is a rather different model, and to say that clarity is "*in* everything, and everything is *in* it" (282) is a very different model indeed. Can you investigate this suggestion in your own experience, along the lines suggested by the text? Notice the statement (283) that the 'clarity dimension' of knowingness is "too subtle for the 'self' to appreciate," and the suggestion that follows for how to break through this limitation.

Chapter 19 of DTS is very direct about the relation of light and knowledge transmission. It specifically grounds this suggestion in nothing more substantial than speculation (199). This is a point we have touched on before, but it seems worth asking again: Can we really expect knowledge to arise through speculation? If so, how could this happen?

Chapter 20 of DTS makes clear that the light that transmits knowledge is not the light that makes 'objects' visible to a 'subject'. This ordinary light creates "a world based on light as it is reflected off the surface of our constructs" and "bathes" experience in a radiating gloom—a "darklight" that leaves everything immersed in shadowy obscurity (209). Can you consider the conventional 'clarity' that gives ordinary experience as 'darklight'? Does it make sense to try?

Much of the DTS reading offers instructions for exploring experience. For example, consider this very precise description:

> The discipline that could lead to . . . lightness . . . starts with fully inhabiting the situation we find ourselves in . . . In the depths of darkness there are traces of clarity. Drawing on moments when this clarity makes itself available, we can refuse the old answers . . . Willingness can see us through obstacles . . . At the 'bottom' of our minds, our senses, and our emotions we can discover the very subtle, sensitive light that purity makes available. DTS 211

Exercises

DTS Ex. 21–22 work with light and darkness directly, complementing the reading, and it would be good to explore them in some depth. In doing Ex. 22D, you could get a sense of what is being suggested by stimulating strong feelings with a Kum Nye exercise related to light; for instance, Ex. 29 (Clear Light) or Ex. 105 (Heart Gold Thread).

One of the two exercises for this week, Appearance Only, is adapted from the DTS reading. Simplifying the text, the exercise could be phrased as follows:

A. As experience arises, let yourself sink 'beneath' the content of mind, turning toward the sense of illumination within which that content appears. As you contact thoughts and feelings at this deeper level, cultivate a clarity that lets you see how you take up each new position.

B. As each position emerges, enter into it, growing intimate with its byways. Look for a sense of illumination within experience, truly unknown, filled with lucent awareness.

C. Instead of abiding within this illumination, set forth again. Open and embody all that you encounter, letting light be your guide. Eventually, there may be a shift in which you become light, and the usual voices fall silent.

For a related exercise, see Awareness Beyond Thought, Ex. 66 in Tarthang Tulku's *Mastering Successful Work,* 202–203, and the discussion that immediately follows.

Unit Thirty-Six
Knowledge and Being

TSK pp. 293–306; DTS pp. 38–41, 156–60, 228–34; KTS pp. 432–35, 477–89

Exercises: TSK Ex. 34; LOK Ex. 50

Supplementary Exercise: TSK Ex. 1–35

Most people find the concluding chapter from TSK, assigned for this unit, deeply inspiring. It suggests that Space, Time and Knowledge shine forth when we learn to appreciate and embody Being, "a mystery when stated in first level terms but quintessentially natural and forthright from a third level view." In the light of being, all three facets of the Time-Space-Knowledge vision are inseparable, and (see TSK 294–95) all the exercises, images, critiques, and qualifications presented as part of the vision serve as "a multi-sided approach to the elicitation of the nature of Being."

The first reading from DTS, previously assigned, introduces the concept of eknosis (see Unit Twenty-Seven). The same concept is presented in a new light in the second DTS reading. Can you get a sense of the way of being evoked here?

Can you explore in your own experience the insight that the "substance and certainty" of conventional experience (DTS 159) can be traced to the operation of underlying mechanisms? One way to go into this possibility would be to apply it to TSK Ex. 34, assigned for this unit, with which we already have extensive experience.

Being is related in interesting ways to the Body of Knowledge:

> The commanding and stalwart presence of Knowledge, its effervescence and youthfulness, the comprehensive intimacy of the *Body of Knowledge*, and the complex, nonaggregated and non-entitative 'totality' all constitute aspects of the full value of appearance and of Being. TSK 293

You may want to explore this relation by doing some searches on the CD, or by working with the passages and exercises in the TSK Guide, Topics 72–73. As a starting point for such inquiry that also points back toward themes we have considered earlier, consider the following passage (DTS 234):

> Just as infinity cannot be divided, the Body of Knowledge cannot be characterized or parceled out; it cannot be acquired or made randomly available.

How does this list of "cannots" relate to the inward-outward knowing of eknosis? Insofar as it names directions, is eknosis still bound to first-level structures? Or can these directions and this naming be understood as symbolic? What do the movement of eknosis and the "nonaggregated" quality of Being have to do with abiding in the zeroless? Finally, how can you take these questions seriously without 'theorizing' about the answers?

The reading from KTS again gathers together the whole of the TSK vision in a way that honors the special precision and inspiration of TSK inquiry. The key idea introduced here is the 'point of decision':

> As a gathering together of all that contributes to its making, each 'point of decision' is the whole. We could even go further: Each 'point of decision' is the emerging gathering of *all known knowledge*, at whatever level it operates. In the gathered wholeness of the 'point of decision', nothing is left out, and choice is freely available. The known world is sustained from moment to moment on the basis of ever-emerging 'points of decision'. *In this ongoing 'emergency', all that is is invariably at stake.* Since this is so, what binds us 'now' is evanescent—freedom can be 'recovered' at any moment. KTS 484

In this program, we have reached the decisive point. The decisive point is the decision point, and the decision point is where we will live the rest of our lives. Has this program helped you live in that 'ongoing emergency' resolutely and without limitation? Has it pointed you in that direction?

Exercises

LOK Ex. 50 is a meditation on the decisive point. The supplementary exercise, below, is meant to embody all that has come before:

> Sit with the TSK book open to the list of exercises (ix–x). Begin going through each exercise, doing each one for a few minutes or even less, following your own sense of what is appropriate. Then move on to the next exercise. If the name of the exercise is not enough to remind you of what it is, skip it and move on to the next.
>
> Continue until you have gone through all the exercises. When you come to Going without Going, you may wish to visualize doing it rather than actually standing up; when you come to Evocation of Knowledge, you may wish to do it silently. Feel free to find your own way.

TSK

Guide

75 Topics

This guide organizes the TSK vision into seventy-five topics, as shown on the next page. The topics are organized in terms of time, space, and knowledge, and within each of those facets of the vision, they are listed in terms of first level, second level, and third level.

This classification system is not meant to be precise, nor is the list of topics meant to be complete. The aim is to give ready access to materials that might otherwise be hard to locate, and also to present a tool that can be of use in investigating the three levels of space, time, and knowledge. Topics not treated in depth in the first TSK book are shown on the list in brackets.

Each of the seventy-five topics is organized into two sections: Analysis and Experience. The section on analysis always begins with a primary reference and then lists other references, citing the books in the order of their publication. The section on experience usually lists exercises from the books, but sometimes cites passages that can easily be adapted as exercises. References are by no means exhaustive: The indexes to each book, or a search through the CD, will provide many other references and connections.

Topics by Division

Space

Level One

1 Appreciation
2 Focal Setting
3 Lower Space
4 Subject and Object
5 Here' and 'There'
6 Polarity and Duality
7 [Projection]

Level Two

8 Body
9 [Senses]
10 Reality
11 Existence
12 Bystander
13 World
14 [Gravity]
15 Partitions and Limits
16 Given Together
17 [Fields]
18 Mind
19 Thoughts
20 [Multidimensionality]

Level Three

21 Symbols
22 [Zero]
23 [Ratio and 16]
24 Space Knowledgeability
25 Great Space

Time

Level One

26 Linear Time
27 [Past, Present, Future]
28 Self
29 Causal Sequence
30 Beginnings
31 ['From/to']
32 Positions
33 [Temporal order]
34 Narrative/story
35 [Founding Story]
36 Feeling and Emotion
37 [Desire]
38 Meaning
39 Going and Friction
40 Momentum

Level Two

41 Timing
42 Point and Pointing
43 [Transition]
44 [Rhythm]
45 Read-out Law
46 [Conducting]
47 The Sacred
48 Dynamic of Time

Level Three

49 [Nuclear Time]
50 Great Time

Knowledge

Level One

51 Lower Knowledge
52 [Technological Knowledge]
53 [Polar Knowledge]
54 Concepts and Beliefs
55 Interpretation
56 Identification
57 Language
58 Inquiry and Analysis

Level Two

59 Not-knowing
60 [X]
61 [Witness]
62 [Tracing]
63 [Imagination]
64 [Speculation]
65 [Logos]
66 Opposition
67 Negation

Level Three

68 Awareness
69 Intimacy
70 Knowingness
71 Light
72 Body of Knowledge
73 Being
74 Love
75 Great Knowledge

Level One
Space

1. Appreciation

Analysis

Primary Reference: TSK pp. 3–10

TSK starts by asking us to reflect on the marvel of space: always available, accommodating all appearance, always leading beyond. Appreciation requires no new understanding; it can be activated at the lower level of space we usually occupy. But it encourages consideration of space, and thus prepares for calling our ordinary understanding of space into question.

Other References
KTS Ch. 25–28
VOK questions 10–15; Ch. 8.3
DTS Ch. 4–5
SDTS Ch. 1:8 (end)

Experience

KTS Ch. 28 (Containment)
DTS Ex. 7

2. Focal Setting

Analysis

Primary Reference: TSK pp. 4–5, 64–68

The ability to see lower-level space as the output of a particular focal setting is the key to arriving at a level-one understanding of space: "our ordinary realm, seen in the light of further possibilities for human fulfillment" (TSK 111). This insight creates the possibility of changing the focal setting and thus gaining access to "qualitatively new spaces." Thus, the focal-setting insight is presented in TSK as a vital method for arriving at an understanding of conventional experience as an output of lower-level space. The methodology involved is described at TSK 64: "This is accomplished by working with what could be called outputs of the 'mind setting' *in ways that are uncharacteristic for that setting.*"

Other References
LOK pp. 414–15 (the relation between focal setting, read-out, and logos)

Experience

TSK Ex. 1–6 (Giant Body)
TSK Ex. 14 (A New Focus on Space)

Almost any of the exercises in the TSK books could be viewed as inviting a shift in focal setting. For an interesting example, see LOK Ex. 19 (Object of Desire).

3. Lower Space

Analysis

Primary Reference: TSK Ch. 1, especially pp. 12–18

Lower space is the realm we inhabit, populated by self and objects, world as background, and so forth. We let ourselves be run by the structures it imposes (see the image of the "petty official," on p. 14). First-level inquiry into space begins with seeing lower space for what it is and accepting the possibility of an alternative.

Other References
KTS Ch. 29–30
VOK questions 1–25
DTS Ch. 4–6
SDTS, "fifteen" (use CD)

Experience

TSK Ex. 1–3 (Giant Body)
TSK Ex. 28 (Transcendence of Pointings)
LOK Ex. 1
KTS Ch. 32 (Silent Partner)

4. Subject and Object

Analysis

Primary Reference: TSK pp. 31–34, 166–74

TSK presents subject/object and self in an integrated manner, although the distinction comes out clearly in associated exercises. Later books tend to treat the subject as an 'entity' distinct from the self. The self, or person, exhibits continuity over time, whereas the subject is simply one pole in any act of subject-object knowing (and thus constitutes a specific instance of polarity). The two approaches lead to a different kind of analysis and different experiential approaches. Both are linked to the topics of positioning and the bystander. See Topics 6 (Polarity and Duality), 28 (Self), 12 (Bystander), 61 (Witness), and 32 (Positions).

Other References
LOK Ch. 13–14; pp. 150–52
KTS Ch. 8, 67
VOK Ch. 3.2
SDTS pp. 17–21, 166–67

Experience

TSK Ex. 9, 17, 25, 26, 30
LOK Ex. 30
KTS Ch. 56 (Co-relation)

5. 'Here' and 'There'

Analysis

Primary Reference: TSK, Ch. 4; Ch. 3, pp. 61–62

'Here' is linked to by-standers and outside-standers: "'Here' is a distracting 'by-stander' in a world of 'outside-standers'" (TSK 74). The sense is that 'here' and 'there' involve us in an endless stream of things and positions. On the other hand, 'here' can be transformed into "the nearness of infinity" (TSK 75; cf. TSK 126). Compare "the vital hereness of the actual" (DTS 89). See Topics 6 (Polarity and Duality) and 22 (Zero).

Other References

LOK Ch. 5 ('here' as private and cut off); Ch. 32;
cf. Ch. 13–14 (polar knowledge)

KTS Ch. 37 (locatedness); Ch. 67 (bystander 'here');
Ch. 85 (projection of 'Here I am')

VOK Ch. 8.1 (We transcend 'here' and 'there' in
being always transient.)

DTS pp. 102–9 ('Here I am' and the witness)

SDTS, Ch. 1:1:("I am here"); also p. 129

Experience

TSK Ex. 25, 26, 29

LOK Ex. 24

DTS Ex. 8

6. Polarity and Duality

Analysis

Primary Reference: TSK pp. 168–74

The structures of polarity and duality are at work in the relation between subject and object, self and world, and 'here' and 'there', all dealt with as separate topics. However, the structure as such, apart from how it manifests, is fundamental to first-level space. It is singled out in TSK, but still in connection with the self and its world. Later books treat it as its own topic, independent of self and world. Compare the related theme of distance and separation (e.g., TSK 61–62), as well as Topic 56: Identification.

Other References
LOK Ch. 13–14; cf. Ch. 24 (polar knowledge);
Ch. 39–40 (negation)
KTS Ch. 57 (polarity and not-knowing);
Ch. 41–44 (negation)
SDTS p. 46 (distance and separation
as knowledge operators)

Experience

TSK Ex. 17 (The Object and Its Glow)
Ex. 26 (Transcending of Pointings)
Ex. 25 (Intimacy)

LOK Ex. 6 (Thoughts in Conflict)
Ex. 39 (Sameness in Difference)

7. Projection

Analysis

Primary Reference
KTS Ch. 85 (Structure of Projection)

The topic of projection is presented in TSK in terms of "space projecting space into space." cf. KTS Ch. 53. Compare this to the kind of projection that sets up lower space, which is the focus here.

Other References
LOK Ch. 32
VOK Ch. 4.2
DTS Ch. 2–3; pp. 150–51
SDTS Ch. 3:2

Experience

TSK Ex. 16 (with special reference to space projecting space)
LOK Ex. 21

Level Two
Space

8. Body

Analysis

Primary Reference: TSK Ch. 2 (Giant Body)

The TSK Giant Body exercises start with first-level space and move to the second-level by opening up and shifting the focal setting. This shift then goes beyond body to 'embodiment'.

Other References
KTS Ch. 28
SDTS pp. 193–94

The Giant Body exercises do not invoke conventional experience as their starting point. KTS Ch. 28 does, inviting us to regard the external world in Giant Body terms. DTS Ex. 7 reverses the direction of Giant Body inquiry.

Experience

TSK Ex. 1–3, 4–6, 7–10, 14
TSK Ex. 34 (a third-level approach, linked to light)
LOK Ex. 3, 24, 38

9. Senses

Analysis

Primary Reference: KTS Ch. 7, 45–48, 55, 68, 82–83, 86–88, 89

Other References

TSK p. 84 (the ontological status of sense data. Many TSK exercises work with the senses.)

LOK Ch. 13 (an analytic, first-level inquiry into how sense-knowledge forms. See also discussions of subject/object relations in all the books.)

VOK Ch. 8.3

Experience

TSK Ex. 7, 17, 24, 29

LOK Ex. 28, 30, 31, 36, 37

KTS Ex. 44, 48

DTS Ex. 6, 9B, 20, 24

10. Reality

Analysis

Primary Reference: TSK Ch. 4

Other References
TSK Ch. 11 (reality as interpretation)
LOK Ch. 17 (reality as interpretation)
LOK Ch. 42–43 (construction of substance)
KTS Ch. 25–28
DTS Ch. 2; pp. 267–74
SDTS Ch. 1.6; pp. 186–88

Topics 10–17 are closely linked, but offer different perspectives.

Experience

TSK Ex. 15D, 15F
KTS Ex. 40
DTS Ex. 7

11. Existence

Analysis

Primary Reference: TSK pp. 31–34, 42–46, 98–108

Other References
LOK Ch. 40
KTS Ch. 103–4; Ch 18 (p. 84: "What 'is' emerges as an echo of itself.")
VOK Questions 52–54
DTS pp. 107–108, 186–88
See also Topics 67 (Negation) and 56 (Identification).

Experience

KTS Ch. 28; exercises under Topic 10

12. Bystander

Analysis

Primary Reference: TSK Ch. 4

The notion of bystander (or 'by-stander') is linked to the notions of subject, self, and witness; all these topics should be consulted. However, the bystander adopts a particular stance toward what is known and experienced—the world of outside-standers. The shift from a focus on self and subject, both items disclosed in the standard world order, to the structure of bystander-outsider reflects the transition from level-one to level-two analysis.

Other References
LOK Ch. 32–33
KTS Ch. 5–8, 37
KTS Ch. 67 (introduces notion of 'bias-stander')
DTS Ch. 17
SDTS pp. 217–28
See also Topics 4 (Subject and Object) and 28 (Self)

Experience

See exercises for Topic 4

13. World

Analysis

Primary Reference: TSK Ch. 4

In contrast to Topics 10 and 11, the topic of world (or world order) deals with "the presupposition of a highly ordered world as the independent and containing background for all things, meanings, and observations . . . , the fundament and touchstone of reality . . ." (TSK 71). TSK could be said to challenge lower-space structures in the following sequence: body, mind, and world.

Other References
LOK pp. 33–36, 102–6; Ch. 18. See also p. 144:
"Because the self 'exists', the world is 'real'."
KTS Ch. 30 (the universe)
SDTS pp. 177–83 (the universe)

Experience

TSK Ex. 15E (A Mountain Retreat)
See also exercises for Topics 10 and 11

14. Gravity

Analysis

Primary Reference: KTS Ch. 15, 21, 38, 85

Gravity constitutes a world or field. It is particularly associated with the momentum of thinking and the 'Here I am', and also with the momentum of projection.

Other References
TSK pp. 237–38
LOK p. 274 (gravity generated within a field)
VOK p. 147 (gravity of thinking)
DTS pp. 53–56
SDTS pp. 206–7

Experience

TSK Ex. 23 (Going without Going)
LOK Ex. 8 (Gravity of Thinking)
DTS Ex. 3 (Playfulness of Thoughts)

15. Partitions and Limits

Analysis

Primary Reference: TSK pp. 28–46, 91–93

Other References

LOK Ch. 30–31. The treatment here calls attention to the limit that establishes the whole, thus arriving at the topic of the field.

KTS Ch. 25–28 (borders)

DTS Ch. 3 (limits); Ch. 6 (edge of the edge) Compare pp. 9–10 (taking apart the box)

SDTS Ch. 1:1 (frame). The discussion here proceeds in terms of the geometry of partition: Every border is a line, and all lines are constructed, starting with the baseline. See Topic 31 (From/To)

Experience

TSK Ex. 5, 8, 9, 14, 15B

KTS Ch. 52 (Lineage of Appearances)

DTS Ex. 7 (Qualities of Space)

16. Given Together

Analysis

Primary Reference: TSK pp. 82–88

Other References

KTS Ch. 41–42

See also the following topics: Fields, Focal Setting, Read-out Law, Ratio, and 16

Experience

TSK Ex. 15E (Mountain Retreat), 26 (Transcendence of Pointings), 27–30

LOK Ex. 16, 25

17. Fields

Analysis

Primary Reference: KTS Ch. 37–44; 83–84
(cf. TSK pp. 97, 158)

Other References
DTS Ch. 2–3 (field and field communiqué); see also
Topics 2 (Focal Setting), 65 (Logos)

Experience

TSK Ex. 15; cf. LOK Ex. 27
LOK Ex. 17, 29, 41, 42
DTS Ex. 5–6; see also exercises for Topic 16

18. Mind

Analysis

Primary Reference: TSK Ch. 3; cf. pp. 288–89

TSK suggests that mind is like space, and that thoughts are like objects in space. In support, it shows a) the structures of space, including such structures as 'here' (TSK 58), apply to any experience, whether the object of that experience is a thought or a sense-object; b) mind, like space, must be accommodating; that "mind, like space, has no foundation" (TSK 64). At the very least, then, mind is *like* space.

Other References
KTS Ch. 7, 33–35, 72, 101, 107
VOK Ch. 6–8
DTS pp. xxix–xxxv; 128–130

Experience

TSK Ex. 7, 11–13
LOK Ex. 2, 40, 45, 46
KTS Ch. 36 (Catching Mind); DTS Ex. 4

19. Thoughts

Analysis

Primary Reference: TSK Ch. 3

On the relation between thought, mind, and space, see TSK pp. 3–4, 14–15, 60–62. On whether there is still a thought when thought stops, see the commentary to Ex. 12, and compare Topic 35 (Founding Story) and Topic 38 (Meaning).

Other References

KTS Ch. 44

VOK questions 43–44; Ch 5.3, 6.2, 7.2, 8.1, 8.2

DTS Ch. 3, 7–8, 20; pp. 245–47, 258–60

Experience

TSK Ex. 7, 11–14

LOK Ex. 4, 6, 9, 39, 43

DTS Ex. 1–6, 9C, 10, 16B

20. Multidimensionality

Analysis

Primary Reference: DTS Ch. 8

Other References

TSK pp. 8–14 (nested spaces), 198–205 (things as "streams of sheer surface appearance")

DTS Ch. 6 (opening all dimensions)

KTS Ch. 45–46, 70

SDTS Ch. 1:3

Experience

LOK Ex. 16

DTS Ex. 7A, 9, 10

TSK Ex. 1–13; compare the momentum of inquiry; cf. DTS p. 66

Level Three
Space

21. Symbols

Analysis

Primary Reference: TSK pp. 9, 11, 14, 144–45

The topic of symbols does not receive extended discussion in any of the TSK books, but the nature of symbols is invoked repeatedly to suggest how appearance can appear without asserting the claims of existence. Symbols themselves can operate in both lower-level and higher-level ways; cf. KTS 475.

Other References
LOK pp. 346–47, 372
SDTS pp. xx, 30–31, 50–51
See also Topics 22 (Zero),
38 (Meaning), 63 (Imagination)

Experience

LOK Ex. 47, 49

22. Zero

Analysis

Primary Reference: LOK Ch. 42, 45

As a symbol for what is not there, zero is inherently paradoxical. The implications are worked out most fully in KTS and SDTS. Cf. TSK 288 (zero-dimensional being).

Other References
KTS Ch. 19–21
Ch. 54, 59, 77–78
DTS Ch. 6–8; pp. 268–69
SDTS pp. 3–5, 30–37, 49–60,
95–100, 120–137, 180–209

KTS adds the notion of the zero point, which later books continue to explore; DTS adds 'the zeroless'. In SDTS, the interplay between the zero of subjective consciousness and the one of objective identity (DTS 50) is expanded and symbolically transfigured in the relation between zero and sixteen. See also Topics 5 ('Here' and 'There') and 42 (Point and Pointing).

Experience

LOK Ex. 27 (Field of Space)
KTS Ch. 36 (Catching Mind)

23. Ratio and 16

Analysis

Primary Reference: SDTS Ch. 1:5; 2:8; 3:6–8
See also Topic 42 (Point and Pointing)

Experience

SDTS pp. 120–23:
Unfolding the Cone; Radiating Cones; Rings in Time

24. Space Knowledgeability

Analysis

Primary Reference: TSK pp. 24–30

The interpenetration of knowledge and space is a hallmark of the TSK vision. See TSK 5 ("we can come to an overall understanding which is itself a kind of space") and 24 ("some kind of knowing is the *primary* fact of embodiment")

Other References
LOK Ch. 26
KTS Ch. 31, 33–36
VOK Ch. 3.3
DTS pp. 38–41; Ch. 6
SDTS Ch. 3:5

Experience

TSK Ex. 4–6, 16
LOK Ex. 43
DTS Ex. 1, 20

25. Great Space

Analysis

Primary Reference: TSK pp. 13–18, 42–46, 64–67; Ch. 4–5

Other References
KTS Ch. 49–56, 95, 103
VOK p. xii
SDTS pp. xxix–xxxiv

Experience

TSK Ex. 4, 10, 12–16, 34
LOK Ex. 46–50
KTS Ch. 52, 56

Level One
Time

26. Linear Time

Analysis

Primary Reference: TSK Ch. 6; pp. 135–41, 178–84

Lower time is linear in its structure. Linear time is linked to meanings (Topic 38) and to the self (Topic 28). It treats time in terms of measurement, as a background.

Other References
LOK Ch. 14–15, 18, 22; KTS Ch. 2, 7–9, 16–18
DTS Ch. 9–10, 13; pp. 156–62; SDTS Ch. 2:2
See also Topics 27–28 and Topic 61 (Witness)

Experience

Many time exercises can be done at the first level to develop awareness of linear structures, and then again at the second level, to transform or open those structures.

TSK Ex. 17–18, 27
LOK Ex. 13–16
DTS Ex. 17–19
SDTS pp. 101–102 (Other Side of Zero)

27. Past, Present, Future

Analysis

Primary Reference: DTS Ch. 9–10; Ch. 11 (pp. 91–94)

Other References
LOK pp. 103–106, 148–50, 186–90
KTS Ch. 1–3, 72

Experience

TSK Ex. 17–22, 27
LOK Ex. 1, 13, 18, 25
DTS Ex. 17–19
SDTS pp. 82–84 (The Feel of the Image),
84–85 (Drawing Closer to Time),
85–86 (Investigating the Future),
86–87 (Edges of Time);
SDTS pp. 214–16 (Alternative Geometries)

28. Self

Analysis

Primary Reference: TSK pp. 30–39, 147–48,
166–74, 178–79

Unlike LOK and the later books, the TSK book does not distinguish between self and subject or self and witness.

Other References

LOK pp. 102–106; Ch. 16–25, 43–44
KTS Ch. 4–6, pp. 65–68, 142–46
VOK pp. 83–84, 106–7, 141–43
DTS, pp. 47–50, 115–25, 142–44, 177–79, 304–6
SDTS pp. 3–14, 17–21
See also Topics 4 (Subject and Object),
12 (Bystander), 18 (Mind), 61 (Witness)

Experience

TSK Ex. 6–10, 17, 18, 25, 30
LOK Ex. 1, 3, 5, 18, 19, 21, 24
DTS Ex. 11–12

29. Causal Sequence

Analysis

Primary Reference: TSK pp. 72–77, 120–23

Other References

LOK, pp. 180–82; KTS Ch. 41–43 ('field totality' as an alternative to causality)

DTS pp. 86–88 (lineage), 139–42; Ch. 18

Experience

TSK Ex. 18, 19, 22A, 24, 26

KTS Ch. 44 (Feel of Space); cf. TSK Ex. 12

VOK question 19

DTS Ex. 18–19

SDTS pp. 102–103 (Cone of History)

30. Beginnings

Analysis

Primary Reference: TSK pp. 72–74, 165–68, 391

In the TSK book, beginnings are discussed in connection with causal sequence (Topic 29). The theme of beginnings is taken up independently in the latter books. As 'the prior', beginnings invoke a higher-order temporality.

Other References

LOK pp. 121–25; cf. p. 116 (each link must have a beginning and ending)

KTS Ch. 19 (first moment as zero; as 'prior'); Ch. 41–43; pp. 352–53

VOK Ch. 5.2

DTS pp. 141–42: the prior; 187–90

SDTS p. xxxii, Ch. 2:4–2:5; pp. 130–34

See also Topics 29 (Causal Sequence), 31 (From/to)

Experience

TSK Ex. 23, 27

See also exercises for Topic 29

31. 'From/to'

Analysis

Primary Reference: SDTS Ch. 1:1; 2:8; 3:2–3

Other References
TSK pp. 59, 104
LOK p. 114; KTS p. 15
VOK pp. 24, 144, 150
DTS pp. 162–64

Experience

TSK Ex. 11, 27
KTS Ch. 52 (Lineage of Appearances)
SDTS pp. 158–59 (From and Not-From)

32. Positions

Analysis

Primary Reference: TSK pp. 49–51

Other References
TSK p. 102
LOK Ch. 32; pp. 132–34, 147–49, 272–75, 379–81
KTS Ch. 73–76; pp. 70–71, 318–21
VOK questions 47–50 (juxtaposition)
DTS pp. 173–76
SDTS pp. 156–58 (juxtaposition)
See also Topic 42 (Point and Pointing)

Experience

TSK Ex. 15D, 17, 25, 28–30
LOK Ex. 5, 24, 28, 30
DTS Ex. 1, 2, 6

33. Temporal Order

Analysis

Primary Reference: LOK pp. 102–6, 188–90

Other References

TSK p. 125

KTS Ch. 8, 10, 18

DTS pp. 125–44, 285–94

See also Topics 32 (Positions), 55 (Interpretation), 65 (Logos)

Experience

TSK 7, 16, 24

LOK Ex. 3, 19, 29, 34

DTS Ex. 11, 12

34. Narrative/story

Analysis

Primary Reference: TSK Ch. 4

The analysis of "world and other bases" in Ch. 4 of TSK uses the notions of 'focal setting' and 'given together' (Topics 2, 16) to challenge our understanding of the physical universe as real. Later books use these and related notions (e.g., temporal order) to extend this challenge to whatever we accept as real.

Other References

LOK Ch. 21–23

KTS Ch. 90

DTS pp. 61–64, 116–19, 285–91

See also Topics 35 (Founding Story), 45 (Read-out Law), 65 (Logos)

Experience

TSK Ex. 1–7, 26; LOK Ex. 1–5, 21

VOK questions 44–49

DTS Ex. 11, 12, 19

35. Founding Story

Analysis

Primary Reference: LOK Ch. 21; pp. 196–98; Ch. 25

Other References
SDTS pp. 5–10, 17–21
KTS Ch. 85
DTS pp. 102–109
See also Topics 28 (Self), 61 (Witness)

Experience

TSK Ex. 23, 25–26, 29
LOK Ex. 5, 24

36. Feeling and Emotion

Analysis

Primary Reference: TSK pp. 263–71

Other References
LOK Ch. 5, 19; pp. 292–94
KTS Ch. 4
DTS pp. 115–25

Experience

TSK Ex. 7, 24
LOK Ex. 4, 6, 17, 20, 35–36, 38–39, 41, 44
KTS Ch. 84 (Fields of Feeling)
DTS Ex. 4D, 6, 9B

37. Desire

Analysis

Primary Reference: LOK Ch. 5, 19

Other References
VOK Ch. 2.2

Experience

LOK Ex. 19, 29, 34

38. Meaning

Analysis

Primary Reference: TSK pp. 97–108

The presentation in TSK emphasizes the ways that meanings (in the sense of "this means that") constitute first-level time. Other readings share this perspective, but also point to "meaning" at a different level, where meaning is what makes experience meaningful. Compare TSK 236; cf. TSK 170–71, 194–95, 203–5, 233–39, 284–85. See in this connection Topics 21 (Symbols), 34 (Narrative/story), and all the topics under Knowledge Level One.

Other References
LOK pp. 141–42, 345–47 (meaning and symbols)
DTS p. 199 (meaning and meaningfulness)
SDTS Ch. 1:5 (ratio)
See also Topics 45 (Read-out Law), 23 (Ratio and 16)

Experience

TSK Ex. 12–13
LOK Ex. 16–17
DTS Ex. 3–4, 17

39. Going and Friction

Analysis

Primary Reference: TSK pp. 100–108, 125, 161–62

The dynamic of time moves from the frozen structures of linear time to an intermediate awareness of going, linked to 'friction' or 'grating' (TSK 103, 161) and to transitoriness. This is closely linked to momentum (Topic 40) and 'timing' (Topic 42) . For the transition to not-going, see TSK 179–84; see also Topic 22 (Zero) and Topic 49 (Nuclear Time).

Other References
LOK Ch. 30 (going, friction, and limits on knowledge)
KTS Ch. 3, 18, 20 (going/continuity as mechanism for establishing); pp. 75, 80 (timed-out)
DTS Ch. 30 (pain of time)

Experience

TSK Ex. 23
LOK Ex. 30
DTS Ex. 5–6, 13–16

40. Momentum

Analysis

Primary Reference: TSK pp. 120–32

The momentum of time can mark a transition, for it points both toward first-level linear time and toward the dynamic of time that leads to the second level.

Other References
LOK Ch. 18; pp. 173–76, 185–90
KTS Ch. 4–6, 16–18, 21 (momentum as transition to second-level time)
DTS Ch. 11 (momentum of time as the aliveness of the future); pp. 285–89
See also Topics 26 (Linear Time), 48 (Dynamic of Time)

Experience

TSK Ex. 18–24
LOK Ex. 3, 7, 13–15, 18, 25–26, 32–33
DTS Ex. 14–16

Level 2
Time

41. Timing

Analysis

Primary Reference: TSK pp. 143–58, 178–84

'Timing' engages the dynamic of time without invoking the 'going' of lower-level time. It can be activated by 'feeding time into time'. (TSK 149–50)

Other References
DTS Ch. 23

Experience

TSK Ex. 18–22
DTS Ex. 13, 16

42. Point and Pointing

Analysis

Primary Reference: TSK pp. 148–58, 178–87

Other References

TSK pp. 301–4 (one point as all points);
cf. LOK p. 398

KTS Ch. 15, 18, 21, 53–54, 59,
107–109 (decisive point)

VOK questions 50–55; pp. xii, 171–75

DTS pp. 65–70, 162–68, 295–302

SDTS pp. 129–37; pp. 199–202 (decisive point);
pp. 222–37 (point as zero)

See also Topics 22 (Zero), 44 (Rhythm)

Experience

TSK Ex. 18–19, 23, 26

LOK Ex. 14, 22–23

DTS Ex. 13

43. Transition

Analysis

Primary Reference: VOK Ch. 8.1–2

Other References

TSK pp. 146–48; DTS pp. 65–66, 96–99, 242

SDTS pp. 36–44, 124–34, Ch. 3:6, pp. 213–27

Experience

TSK Ex. 22–26

LOK Ex. 22–23

44. Rhythm

Analysis

Primary Reference: KTS Ch. 12–20, 43

Other References
LOK pp. 398–400
VOK questions 26–30
DTS pp. 145–48
See also Topics 42 (Points and Pointing), 48 (Dynamic of Time)

Experience

TSK Ex. 24–26
LOK Ex. 15, 26; LOK 19, 29
LOK 9, 10; LOK 31, 37
DTS Ex. 13–16

45. Read-out Law

Analysis

Primary Reference: TSK Ch. 9

Other References
LOK Ch. 38, 45; p. 415
KTS Ch. 4, 14, 97; pp. 106–8
VOK pp. 55–56; Ch. 4.1
See also Topics 33–35, 65 (Logos)

Experience

TSK Ex. 26, 29
LOK Ex. 41–43
DTS Ex. 11

46. Conducting

Analysis

Primary Reference: DTS Ch. 13–15, 31;
cf. KTS pp. 246, 267, 367

Experience

DTS Ex. 4–6, 17–19

47. The Sacred

Analysis

Primary Reference: TSK pp. 75–77, 157–58, 202–205

Other References
LOK Ch. 28–29; KTS Ch. 24, 76
VOK pp. 165–67; SDTS pp. 150–53

Experience

TSK Ex. 32, 33
LOK Ex. 34; LOK Ex. 6, 39, 40, 42, 45, 46

48. Dynamic of Time

Analysis

Primary Reference: TSK pp. 106–108; 143–58

Other References
LOK Ch. 49; KTS Ch. 15, 16–18
DTS Ch. 11; SDTS Ch. 1:7

Experience

TSK Ex. 23–26, 34
DTS Ex. 11–13, 14–16
SDTS pp. 86–87 (Edges of Time)

Level Three
Time

49. Nuclear Time

Analysis

Primary Reference: DTS Ch. 15–16

Differently dimensioned, nuclear time allows for 'finity' and the universal unique.

Other References
DTS p. 231; KTS p. 296
TSK pp. 179–84 (the transition from 'going' to centered abiding and the centerless center)
VOK Ch. 4.3; pp. 171–75 (universal unique)
SDTS pp. 136–37 (finity), 196–98 (dynamics of 16); 229–30 (finity and 16)
See also Topic 45 (Read-out Law)

Experience

DTS Ex. 13–14; TSK Ex. 23

50. Great Time

Analysis

Primary Reference: TSK pp. 98–110, 159–62

Other References
LOK Ch. 49
KTS Ch. 19–24, 95, 103
DTS Ch. 16 (Body of Time); pp. 228–34
(Body of Energy; universal unique)
SDTS pp. xxix–xxxiv

Experience

TSK Ex. 24–26, 34–35
LOK Ex. 46–50
DTS Ex. 16

Level One
Knowledge

51. Lower Knowledge

Analysis

Primary Reference: TSK Ch. 12

The TSK critique of lower knowledge is extended in later books, especially LOK. The image of the cone (TSK 246) becomes a central organizing principle in SDTS; cf. DTS 131–33.

Other References

LOK Ch. 13–20 (polar knowledge, descriptive knowledge, temporal knowledge, collective knowledge, models);
LOK p. 353 (lower knowledge as knowledge)

KTS Ch. 10–12, 65–67

DTS Ch. 13; pp. 102–5, 184–86

See also Topics 52–58

Experience

TSK Ex. 27–29

LOK Ex. 2–5, 16, 19, 29

DTS Ex. 1–2

SDTS, p. 120–23 (Unfolding the Cone, Radiating Cones, Rings in Time)

52. Technological Knowledge

Analysis

Primary Reference: LOK Ch. 3–5; pp. 158–160

Other References
TSK pp. 113–18 (control and improvement orientation); 137–38 (lower time and technology)
VOK 2.1–2

Experience

TSK Ex. 16
LOK Ex. 19, 29
See also exercises for Topic 51

53. Polar Knowledge

Analysis

Primary Reference: LOK Ch. 13–14; pp. 194–96

Other References
TSK pp. 167–71
KTS Ch. 57, 85 (Structure of Projection)
See also Topics 4 (Subject and Object) and 6 (Polarity and Duality)

Experience

TSK Ex. 9, 17, 25–26, 30
LOK Ex. 30
KTS Ch. 56 (Co-Relation)
SDTS p. 154 (Cone of Knowledge)

54. Concepts and Beliefs

Analysis

Primary Reference: TSK Ch. 11,
Ch. 12 (pp. 233–39)

Other References
LOK Ch. 7, 16; pp. 325–26, 347–48
KTS Ch. 79–81, 104
VOK pp. 3–6, 58–60; questions 46–47, 58
DTS Ch. 24, pp. 267–71
SDTS pp. xxv–xxviii

Experience

TSK Ex. 27–28
LOK Ex. 16, 30, 41; LOK Ex. 6, 39
DTS Ex. 7, 9

55. Interpretation

Analysis

Primary Reference: TSK Ch. 11–12

Other References
LOK Ch. 17–18
KTS Ch. 10
VOK questions 48–50; Ch. 3.2

Experience

LOK Ex. 18, 21
See also exercises for Topic 54

56. Identification

Analysis

Primary Reference: TSK Ch. 4; pp. 31–46

Other References
LOK Ch. 13
KTS Ch. 53
VOK questions 16–17
DTS Ch. 2, 4; pp. 85–90, 145–48

Experience

TSK Ex. 1–10
LOK Ex. 5, 16, 19; 6, 39, 40
DTS Ex. 9–10

57. Language

Analysis

Primary Reference: TSK pp. 98–104, 263–68

The TSK book takes up the topic of language primarily in terms of meaning: "This means that." See Topic 38. Later books analyze language from a variety of perspectives.

Other References
LOK Ch. 6; pp. 141–42, 270–73, 345–48; Ch. 43
KTS Ch. 10, pp. 77–81, 154–55
VOK questions 49–60; pp. 61–63
DTS Ch. 3
SDTS 3:3

Experience

TSK Ex. 13, 26, 28
LOK Ex. 1

58. Inquiry and Analysis

Analysis

Primary Reference: TSK pp. xxxiv, 140–42

TSK proceeds by way of inquiry, and inquiry can be the bridge from one level to the next. As an example of inquiry in action, consider the exercises in TSK Ch. 2. The TSK book emphasizes different kinds of inquiry in operation at different levels, but the picture presented in later books suggests a more complex interaction.

Other References
LOK pp. xix–xxx; Ch. 33–37, 45–47;
pp. 227–28, 352–54
KTS Ch. 61–64
VOK questions 32–37, 43–44; Ch. 1, 5; pp. 113–17;
Ch. 7.2–3; pp. 169–71
DTS pp. xxviii–xxxiii, 65–68, 203–5 (multiple
dimensionality of inquiry),
258–60 (arising of thought)
SDTS pp. xix–xxxix
See also Topic 62 (Tracing)

Experience

TSK Ex. 1–10
LOK Ex. 47–48
DTS Ex. 4, 17–18

Level Two
Knowledge

59. Not-knowing

Analysis

Primary Reference: TSK p. 260; cf. pp. 153–58

Other References

LOK pp. 223–26, 270–72; Ch. 47–48

KTS Ch. 57–60 (not-knowing leads to knowledgeability), 73–76, 94

VOK Ch. 3.2, 9; cf pp. 83–90 (intrinsic knowledgeability)

DTS pp. 102–5; Ch. 14

SDTS pp. 106–7

Experience

TSK Ex. 30–31

LOK Ex. 6, 39, 40; LOK Ex. 11, 16–17, 23, 38

DTS Ex. 1–2, 23

60. X

Analysis

Primary Reference: KTS Ch. 58–59

Other References
SDTS pp. 21–24, 197–98

Experience

LOK Ex. 22–23

61. Witness

Analysis

Primary Reference: LOK Ch. 17; pp. 171–73, 188–90; Ch. 24; pp. 383–84

Other References
KTS Ch. 74–75
DTS Ch. 12
See also Topics 10 (Reality) 12 (Bystander), 28 (Self)

Experience

TSK Ex. 17, 25–26
LOK Ex. 22–23; LOK Ex. 15, 26, 32
DTS Ex. 4–5

62. Tracing

Analysis

Primary Reference: LOK Ch. 33–37

The term 'tracing' as a technical concept seems to appears just once in TSK (41–42), but the methodology figures in many of the exercises. LOK sets out the practice of tracing as "free and open inquiry."

Other References
KTS pp. 92, 237, 274, 402, 432, 477, 486
VOK questions 8, 55

Experience

TSK Ex. 7–9, 18–22
KTS Ex. 52
DTS Ex. 4, 17–18

63. Imagination

Analysis

Primary Reference: VOK pp. 6–13, 58–61

The TSK book, introducing a *vision of reality*, discounts imagination as being only subjective (TSK 26; cf. 204; cf. KTS 467 and VOK 101). Later books, aimed at *implementing knowledge*, treat imagination as a gateway to knowledge. A symbolic turning point can be found in LOK, 89: "Imagination 'learns' to imagine the unimaginable."

Other References
DTS pp. 62–68, 196, 240–42
SDTS p. xxii; Ch. 2:2; cf. LOK pp. 142–144
See also Topics 21 (Symbol), 61 (Witness), 64 (Speculation)

Experience

TSK Ex. 32–34; DTS Ex. 25

64. Speculation

Analysis

Primary Reference: VOK pp. 58–60, 100, 123

Other References
DTS pp. 64–66, 133–135; KTS p. 353 (but see 379)
SDTS pp. xxii, xxv–xxviii, 11–14, 52–56

Experience

DTS Ex. 12B, 17, 18B, 19

65. Logos

Analysis

Primary Reference: LOK pp. 226–28;
Ch. 38, 40; p. 415

Other References
KTS Ch. 4, 8, 10–11, 16–19, 92–94
VOK Ch. 9.2
DTS pp. 155, 193–95
SDTS pp. xxviii–xxix, 50–51

Experience

TSK Ex. 26, 29
LOK Ex. 41–42, 47
DTS Ex. 9, 11

66. Opposition

Analysis

Primary Reference: TSK pp. 71–75, 87–88

The TSK book identifies 'outside-standers' as the entities of our world with which we can interact, and challenges them on the basis of a 'no source from' insight. In later books, these themes are explored in their knowledge aspect in terms of opposition (and distinction), which are presented as the foundation for whatever appears separate.

Other References
LOK pp. 244–46, 264–66; Ch. 39, 42
KTS Ch. 37 (bystander/outstander, opposition and juxtaposition), 74–76, 81–84
VOK question 47; Ch. 3.1, 4.3, 9.2
DTS pp. 146–48, 173–79
SDTS pp. 55–56

Experience

TSK Ex. 9B, 15–16, 25–26, 29–31
LOK Ex. 39
KTS Ch. 84 (Fields of Feeling)

67. Negation

Analysis

Primary Reference: TSK pp. 102–8

In the context of discussing the dichotomy between existence and non-existence, TSK introduces the possibilities of a negation not dependent on duality.

Other References
LOK Ch. 40–45, 50 (negation and the 'ness' of 'ness')
KTS Ch. 41–43, 50, 74–76
DTS pp. 5–8, 30–32
SDTS pp. 224–27
See also Topic 22 (Zero)

Experience

LOK Ex. 40–45
KTS Ch. 44 (Field of Space)
DTS Ex. 1, 5–6

Level Three
Knowledge

68. Awareness

Analysis

Primary Reference: TSK p. 44, 168, 187, 257, 298

Awareness is variable. It can be owned by a self, but it can also open, relax, appreciate, and allow. In that case, it counters the usual "control and improvement orientation" (TSK 118), eases the transition to "doing nothing" (TSK 107, KTS 380), and deepens toward 'knowingness'.

Other References
TSK pp. 76–79 (religious surrender)
LOK Ch. 47; cf. Ch. 29
KTS Ch. 68, 72, 86
See also Topic 70 (Knowingness)

Experience

TSK Ex. 6, 15–16, 23, 29, 32–33
LOK Ex. 10, 47
DTS Ex. 1–3

69. Intimacy

Analysis

Primary Reference: TSK pp. 153–62

Other References
LOK pp. 307–10
KTS Ch. 20, 37; pp. 110–11
VOK pp. 55–58
DTS Ch. 23

Experience

TSK Ex. 24–26
LOK Ex. 37
KTS Ex. 86 (p. 388)
DTS Ex. 16

70. Knowingness

Analysis

Primary Reference: TSK Ch. 14–15

Other References
TSK pp. 159–62, 165–71, 211–14, 219–21
LOK pp. xlv–xlvi, 363–65
KTS Ch. 63–64, 98–99
DTS pp. 138–39

Experience

TSK Ex. 28–35
LOK Ex. 45–46
DTS Ex. 23–24

71. Light

Analysis

Primary Reference: TSK pp. 271–286

Other References
VOK pp. 136–40; KTS Ch. 86, 102–3, 109–12
DTS Ch. 19–21
SDTS pp. 222–24

Experience

TSK Ex. 4, 32–34
LOK Ex. 36–44
DTS Ex. 10, 20–22; pp. 212–14 (Appearance Only)
SDTS p. 223: (Balance Point of Light)
KTS pp. 388–89 (Free Arising)

72. Body of Knowledge

Analysis

Primary Reference: TSK pp. 253–55, 286–90

Other References
LOK Ch. 50
KTS Ch. 98–102; pp. 350–51
VOK Ch. 3.1, 6.1; Conclusion
DTS Ch. 17

Experience

TSK Ex. 34; TSK Ex. 24, 35
LOK Ex. 50
VOK p. 143 (Trinity of Inquiry)
DTS Ex. 6, 24

73. Being

Analysis

Primary Reference: TSK pp. 293–300

Other References
LOK Ch. 45, 49–50
KTS Ch. 19, 96, 100–103
SDTS pp. 231–37

Experience

TSK Ex. 1; LOK Ex. 1; DTS Ex. 1
LOK Ex. 2, 49

74. Love

Analysis

Primary Reference: TSK pp. 263–71

Other References
LOK Ch. 37–38; pp. 409–10
KTS Ch. 112
DTS pp. 333–39

Experience

TSK Ex. 25, 31
KTS Ex. 50
DTS Ex. 24

75. Great Knowledge

Analysis

Primary Reference: TSK pp. 198–205; Ch. 10;
pp. 239–48; Ch. 13–14

Other References
KTS Ch. 51, 55, 61, 95–97, 102–3
SDTS pp. xxix–xxxiv, 235–37

Experience

TSK Ex. 4, 10, 13, 34–35
LOK Ex. 46–50
DTS Ex. 23–25

Nine Levels in Six Books

The presentation that follows analyzes how the three levels of Time, Space, and Knowledge are presented in the books of the TSK vision.

How this happens is very different in each book. The first TSK book is the most systematic in its presentation of the vision in terms of a progression from level one through level three, but superimposes this structure on a development that moves from Space to Time to Knowledge. LOK spends considerable time first presenting and then critiquing level-one forms of time, space, and knowledge; however, in Part Three of LOK there is a jump directly from level-one knowledge to level-three knowledge, via an analysis of the limits that shape conventional knowledge.

KTS could be understood as starting right in with a second-level critique of first-level time. Similarly, SDTS could be seen as grounded from the outset in a third-level critique that is also an evocation and appreciation.

Of course, it is not necessarily important, or even helpful, to develop a 'theory' of how each TSK book unfolds or how the books are interconnected. Instead, the aim is to let the presentation here stimulate new inquiry, new approaches, and new connections.

Level One
Space

Time, Space, and Knowledge

TSK 3–12

Space is opposed to the objects within it, and seems itself untrackable. But this dichotomy may trace to a fundamental unity. Space and objects are bound to one another, and even understanding partakes of space. So we can entertain the idea that space is projecting space into space. This seems an accurate description, even if it departs from our usual understanding. Taking this approach, learning to shift our focal setting, allows for the possibility of opening to Great Space.

TSK 21–26; Ex. 1–3; cf. Ex. 7

Entering the Giant Body, we find space where we previously found only substance, and begin to develop ways to replace the isolated appearances of 'body', 'mind', and 'world'.

TSK 49–58; Ex. 11

In our own experience, we can identify space in contrast to the 'mindings' with which we are familiar: first as the source of thoughts, and then within thoughts. Letting go of the idea of 'source-from' opens the possibility of discovering a different, more spacious space.

TSK 71–79

Casting reality in terms of bystander and outside-stander yields cause-and-effect sequences, including first cause. But Great Space reveals that none of this is necessary. We can clear away the 'familiar world order' and 'here' orientations, together with 'goal and progress' orientations.

Love of Knowledge

Chapters 13–14, 21, 26

Polar and descriptive knowledge work together to give 'objective' reality. In such a reality, space itself, in its allowing, disappears in favor of objects (cf. the beginning of Ch. 21). The founding story and the ongoing narrative confirm objective existence as persisting.

Knowledge of Time and Space

Chapters 25–30

Space is the silent partner that allows. We ignore it, but we can challenge this ignoring; e.g., by looking at the edges between matter and space. When we do, we find ourselves asking about the existence of space, as the 'not' of what is. As a hypothesis, we could imagine objects as projections of space; as blueprints for space that allow us to communicate with space. This is not just theory, for if objects are space, they are open and indeterminate, allowing for a certain freedom.

The relation between space and what occupies it is problematic. For instance, there is a sense in which as individual human beings we always occupy the same space; there is another sense in which nothing ever occupies the same space for more than a moment. Such considerations prepare the way for a second-level view of space.

Chapters 33–34

There is a first-level way to consider the spatiality of mind, investigating the various justifications for such a view, the difficulties that denying it raises, and the new possibilities it might offer. All this clears the way for a more straightforward integration of first-level mental and physical space, one that operates from a level-two perspective.

Dynamics of Time and Space

Chapter 1

If we expand matter, it moves toward nothing. If we expand space, what does it move toward? Does it become less than nothing, or perhaps a more 'nothing-full' nothing? Or does the negation of space-nothingness create something positive, something that could be understood as a different kind of space? The question opens space to inquiry.

Chapter 2

Trying to get at space, we look toward substance. But substance seems to depend on interaction; in the end, on the co-referring that constitutes the field communiqué.

Sacred Dimensions of Time and Space

Introduction

Space as the background of our being. Opening space through opening 'I am here'.

Time 3

Every point is a point in both space and time. The two seem perfectly synchronized, connected in complete intimacy.

Level One
Time

Time, Space, and Knowledge

TSK 102–10; Ex. 16

Time is thought of as a linear tube in which static entities appear, giving rise to a stream of meanings based on dichotomies and manifesting in transitoriness. But doing, going, and happening are outcomes of one particular take on Great Time, linked to the picture of a knowing self.

TSK 117–32

Lived time is an abstraction, but also has a relentless power. It is in some ways a hidden force. And nothing at the first level can account for the fact that it presents serially. Unless we work directly with time, we will continue to 'time ourselves away', experiencing tremendous suffering. But each presentation bears the possibility of contacting Great Time.

TSK 135–41

Lower time is the attempted taming and appropriation of Great Time for egoistic purposes. We can challenge it at the first level through deeply thinking it through and penetrating its manifestations.

Love of Knowledge

Chapters 14–15

The structures of linear time cannot sustain the truth of knowledge or description; they depend on a mechanism of linking that breaks down under analysis. The real guarantor is the self. But the self is a creature of time, insisting on its persistence and on its ownership of experience.

Chapters 18–19

The self lives toward the future, and so is cut off from the present. It relies on intentional knowledge, bringing feelings and emotions to the fore, or retreats into technological knowledge. Time is measured out in advance. "The force of the self's desire unfolds as the momentum of linear time."

Chapters 21–23

The self relies on narrative, telling the founding story of the narrator. It asserts two contradictory understandings of time: present as dividing line and as the whole of experience. The relation of self to the flow of time, and the nature of that flow, is problematic, but the flow does suggest something about the dynamic of time; through transition we may gain access to a different understanding.

Knowledge of Time and Space

Chapters 1–6

The structures of linear time—past-present-future, the self and its experience, inner experience and outer objectivity—are mechanisms. They could operate differently.

Chapters 10–11

First-level time controls and limits knowledge, confining it to interpretation and substance.

Chapter 14

The first-level approach to time as a grid for measurement reflects a second-level temporal rhythm.

Dynamics of Time and Space

Chapter 9

We know only the present, not the past and future. And in fact, we do not even know the present, since we have conflicting models of it. Do we know time at all, when we know only its passing away?

Chapter 10

As history that can be identified, the past is inherently limited. For linear time, specifying moments, all time suffers from this limitation. Even the future has 'happeneded'. We live in lifeless ways.

Chapter 13

We conduct all the limitations of our ordinary way of being in conducting linear time. Its structures—sameness and transition, conditioning and identity—call for ongoing interpretive efforts that give the world we know.

Chapters 29–30

We encounter time only in our alienation from it. But we can touch the dynamic of time by letting ourselves feel the pain of time lost, and can balance that pain through an analysis of linear time as untenable.

Sacred Dimensions of Time and Space

Time 1–2

We fail to notice the dynamic of time or inquire into its power; we do not inquire as to how knowledge might unfold differently in time. But we could ask about time's dynamic, about what makes past past and future future, and how transitions are possible.

Time 3–4

We could measure the flow of time, and question its structure differently. Why not many time lines; many directions of flow? Why not time before the starting point?

Time 5

Instead of looking for beginnings, we can look at the 'now' and at the self as it is timed out across time. Here too there are questions and unknowns that let us take a different angle.

Level One
Knowledge

Time, Space, and Knowledge

TSK 221–34

Examining how all ordinary knowledge amounts to variations on a theme can help us break free of its limits. Roughly speaking, it is all 'minding', all images and concepts. Through all of history, we have relied on beliefs. This 'minding and belief structure' trend does not allow criticisms of itself to be made.

TSK 263–71

Lower-level knowing, with its emphasis on labeling, leads into emotionality, setting up negative levels even in the most positive experiences.

Love of Knowledge

Chapters 1–12

Lower knowledge has been set up to depend on beliefs, to cut desire off from knowledge and the subjective from the real.

But we can appreciate the knowledge available to us; we can challenge the limits and patterns that shape us.

Chapters 13–15

The structures of lower knowledge: polar, temporal, and descriptive knowledge. Polar knowledge is fundamental to lower knowledge, but the polar structure in inherently untenable (cf. Ch. 32), due in part to the structure of lower time.

Chapters 16–23

Knowledge that depends on constructs and models may not be knowledge at all. It cannot prove its claims. There is only the witness. This leads to a critique of the self in its relation to linear time. The self presents itself as knower, and thus restricts knowledge. All knowing presupposes the flow of self-centered experience, which can therefore not itself be known.

Chapters 24–26

The self can be challenged as the shadow of its own way of knowing. The witness bears witness to something deeper. For the sake of the identity and existence of the self, we bargain knowledge away, but knowledge could be different, because time and space could be different. While this is just what the self denies, its fear and confusion could be pointers, suggesting ways to challenge the logos.

Chapters 30–32

Tracing level-one limits can be the bridge that leads to level three. But to trace their structure, we must acknowledge their operation. The model of the bystander as fundamental.

Chapter 39

The usual structures of negation, opposition, and substance on which lower knowledge depends. A critique of these structures opens second-level space.

Knowledge of Time and Space

Chapters 10–12

First-level temporality allows knowledge to develop only in terms of interpretation, but second-level temporality opens this up. In turn, even first-level inquiry into time (for instance, the rhythms and cycles of history) can make the dynamic of time more available.

Chapter 57

Not-knowing is fundamental and primordial in lower-level knowledge. Problems and solutions are polarized, and it requires tremendous effort to bring them together.

Chapters 62–63

A critique of first-level knowledge that discovers the second-level knowledge inherent in the limiting and confining order can become the expression of third-level knowledge.

Chapters 65–68

We limit knowledge by assigning it to the bystander-self. But the self as such remains unknown: a dimension of knowledge is hidden. When we turn from the self, that dimension opens.

Chapters 73–75

A 'nothing to say' underlies each position, meaning that in the end no position can affirm what it maintains. Each affirmative depends on a negation.

Visions of Knowledge

Chapters 2–4

The limits of technological knowing and the possibility of inventing a new mind. Technology brings what is new, and right there, in the heart of the new, are the transitions that can open to knowledge. Caught in the fabricated world of conventional knowledge, "we can only dance around the edges of knowledge, anxiously weaving rumors into simulations of knowledgeability." But the gaps and transitions reveal projection. The medium of fabrication is only names.

Chapter 7

We stay on the surface of the mind and miss its vastness. But this self-imposed restriction offers few benefits. If we stay on the surface, we do psychology or science; if we go within, we do philosophy or religion. But the aim is to activate the vastness of mind *within* the operations of ordinary mind.

Dynamics of Time and Space

Chapter 17

Bystander knowledge is set up in such a way that we can never go beyond it. But when we set aside judgments and give up the burden of being knowledge-holder, an intrinsic knowledgeability is there.

Chapter 24

We have sought knowledge, but in a way that pushes knowledge away. We follow our models, but we can let being be the model, so that the Body of Knowledge can be born.

Sacred Dimensions of Time and Space

Introduction

Bringing knowledge to space by starting from the 'I am here', we can see and open the usual space and the usual understandings, and we can penetrate the usual distinctions between subjective and objective.

Knowledge 1

We know in a way that makes time and space obstacles to knowledge. If we get clear on the structure of the cone of knowledge, we open the possibility that knowledge could instead take on a time-space embodiment.

Knowledge 2

Baseline knowledge sets up limits on knowledge; at the same time it cuts us off from space and time. But we could challenge it by not accepting the baseline model for knowledge, either in its history or its moving outward to know.

Knowledge 3

The self insists on distance, which results in limits on knowledge. Language supports this move by creating its own distance. But language can also open new possibilities.

Level Two
Space

Time, Space, and Knowledge

TSK 26–46; Ex. 4–6, 7–10

We can invite more knowing into appearance, entering a world in which there are not really things, but only consolidating tendencies. By looking at the observer, opening all appearances, and noticing 'coming out' tendencies, we can discover openness and trace the knowing capacity active within 'space'.

TSK 58–68; Ex. 12–14

In a higher space dimension, the self and 'from-to' distributions have no special role to play. Initially we need to rely on the outputs of mind, but we can work with them in ways uncharacteristic of the focal setting, and thus open the second level. First-level entities such as the self are seen as tendencies that need not consolidate or 'come out'.

TSK 80–88, 91–98; Ex. 15

We can challenge and open up conventional perspectives in ways that do not establish anything, looking for subtle instances of bystander and outside-stander. We can transparentize partitions, so that we learn to see the output of a focal setting *as* an output. As we allow more space, the challenge can be experientially grounded, and the challenging itself can become a path.

Love of Knowledge

Chapter 48

Opening all limits. Cf. Ch. 39–42.

Knowledge of Time and Space

Chapters 31–32

We can identify a second level of space that allows objects and their surrounding and containing space. This prospect invites more luminosity and more knowledge: a knowledge that may not be ours. We can investigate this experientially, for we learn about space whenever we look to the properties it allows.

Chapters 35–36

From a second-level perspective, first-level knowing and first-level space are naturally given together. Knowing becomes inherent to space, giving to all appearance a special glow or shimmer. We can see how the focal setting is carved out, the angle set up; and we can explore directly the first-level gap between knower and known, which appears as well as the gap between subject and object.

Chapters 37–40

First-level space and its contents are given together as exhibitions of a field. But the field continues to embody allowing, and thus allows for a knowledge that can know field dynamics and mechanics, opening to greater knowledge and to different, even miraculous, kinds of exhibitions.

Chapters 41–48

The field, though bound to first-level entitativeness, opens toward allowing, especially in the focus on transition and change. Self-constructing, it can also be challenged, for its claims to authenticity are internal. Its operation is best understood as a 'not preventing'. All this can be investigated in the realm of the senses, including mind, through which the rhythm of appearance can once more manifest. (cf. p. 377)

Dynamics of Time and Space

Chapter 3

Looking for substance in appearance, we find only the telling of the field communiqué. 'What is' appears as the tracing of zero. The freedom of that zero depends on opening to space.

Chapter 4

We have made space disappear, but we can discover it in the allowing of appearance. Appearance can be seen as space in operation, indeterminate and non-established.

Chapter 5

How can we contact 'no-body space', active within appearance? We can imagine space projecting images inward to the mind, and also imagine mind projecting outward, without losing contact with space inwardness: projecting the inward-outward of eknosis, allowing for space knowledge.

Chapter 6

Flattening space, so that each point is zeroless, gives appearance without substance. The claims of substance and of the knower of substance can no longer be sustained, and the split between the 'one' of substance and the 'zero' of subjective knowing has never happened.

Chapter 7

Thoughts proliferate as the structures perpetuating the substance of what is so. Like space, they appear in not appearing, as no-thing. But we can abide in thought, making it the agent of the zeroless.

Chapter 8

Substance is told by stories, and once told, turns stories themselves into the zeroless. But multiple dimensionality restores the zeroless and opens into new knowledge.

Chapter 27

Space as nothing is unknowable, but when reality becomes multiple and transitional in its arising, space becomes accessible as generative source, not excluded by the claims of substance.

Sacred Dimensions of Time and Space

Space 1–3

If we take 'I am here' as form rather than content, we are led to the baseline and axes; from the zero-point we are led to 16. If we trace the 'here' in the zero-point, we arrive at a fuller, more three-dimensional 16. Cones radiate out, opening the possibility of form and meaningful structure without departing from the indeterminacy of the zero point.

Space 5–7

Space gives rise to the specifics of form through the ratio, which knowledge controls. It allows all appearance, allows knowledge, and allows the unfolding of our destiny.

Knowledge 5

Zero gives access to second-level space, the space that gives conventional space and all that that space allows.

Level Two
Time

Time, Space, and Knowledge

TSK 142–58

Second-level time offers the opportunity of experiencing time directly, something not possible at the first level. It reveals itself in a flashing, dynamic way. We can learn to appreciate immediate presence, a 'timing' in which all manifestations are interrelated. A strong sense of no-self can emerge, while first-level objects and entities are seen as a special kind of timing. Discontinuties and other non-standard or miraculous events may become available. We discover timing to be a dynamic that does not necessarily go anywhere. Time presents infinite possibilities and universal accessibility.

TSK 165–90; Ex. 17–25

The usual first-level stories can be retold as a presentation by time, in which the self eventually emerges, laying claim to the power of time but ceding that power to physical processes as well. This emphasis on the self can be de-emphasized, as can the usual past-present-future structures. We discover the possibility of experiencing moments within moments (though this still concedes too much to first-level views). Eventually we discover within time's dynamic a non-going that merges into intimacy.

TSK 193–208; Ex. 26

Our usual encounters require us to ignore time, but if we experience a second-level 'timing out' of experience, we recognize each situation as complete, a read-out with no relation to any other read-out. Temporal connectedness and similar phenomena are given within the read-out, as is sameness. There is no place, no sequence, no one to be trapped. It is all a play of time.

Love of Knowledge

Chapters 23–25

The dynamic of time does not square with an understanding of time as linear. The witness may give access, and so may the vitality of experience.

Chapter 29

The mystical vision (the vision of the robe) offers time as timeless, as random. But we must be careful: if in moving to this second-level understanding of time we reject ordinary time and conventional reality, we only confirm their limits.

Chapter 33

The relation between the dynamic of time and the dynamic of inquiry. (cf. Ch. 38)

Chapters 48–50

The more we activate knowledge, the more the dynamic of time becomes available. The more we appreciate what time presents, the more time deepens.

Knowledge of Time and Space

Chapters 1–3; 4–6

The first-level structures of linear time are particular mechanisms; it is possible to imagine they could work differently. We place the self at the center, but we may not need to. Time could operate differently; the mind could contact it more directly. But for this to happen, we must challenge the linear, one-dimensional model.

Chapters 7–9

Linear momentum and conventional mind are patternings of the logos, communicating forward the dynamic of second-level time. We could be aware more directly of this dynamic, the force that gives the temporal order, but we have chosen a way of being that cuts off the possibility of such awareness, based on the self's need to exist.

Chapters 10–12

Second-level time allows knowledge to arise as appreciation, beyond language and interpretation. The reliance on substance is loosened, and the dynamic of rhythm becomes available.

Chapters 13–15

Knowledge is given by time's momentum and rhythm: second-level knowledge is found not in substance, but in transition. The insistence on substance obscures this, turning rhythm into points that can be specified and used to identify objects. But such knowledge has no access to the source of time's momentum, available only as the never attainable zero-point approached in looking at points between points.

Chapters 16–18

The momentum of second-level time is the creativity of life itself, accessible within even the most frozen order. In this momentum, there are no entities, only interaction. From rhythm, form takes form, arising as the echo of itself. Friction expresses the demand to exist, and the logos gives rise to the rules that allow for measuring out.

Chapter 35

Second-level knowledge reveals that the temporal order sets up the gap between knower and known, declaring knowledge to be unavailable. (cf. 376–77)

Visions of Knowledge

Chapters 2–3

When we let time come forward, experience is infused with time, and this means that when we investigate experience, we touch time. A key is to discover the lineages of the past and their impact.

Chapter 8

The capacity for sensing that arises in time cannot be traced back to a first moment, nor can we establish with certainty that time has a forward directionality, or that this is the best way to link knowledge and time.

Dynamics of Time and Space

Chapter 11

The future gives us time's momentum. We can return to the dynamic of our own being and invite a boundless knowledge, transmitted in transition, dimensionless and alive.

Chapter 12

With no real access to time's dynamic, we try to impose order by means of the witness. We invoke an order, but do so at the cost of confining knowledge to the present moment. But we do not need substance and identity and the order of 'from/to'. We can turn inward to discover time's dynamic.

Chapter 13

In conducting what appears in time, we also conduct time. Mind does the conducting, but mind is also the conducted. In this circularity, we can sense the openness of the future, free of limits.

Chapter 16

Observing the recordings of time, seeing how recordings point only toward each other, we can discover the mechanism of this pointing.

Chapter 23

An opening inward of time, going beneath the arrow of time, can open the vastness of the Body of Knowledge.

Chapter 26

Exploring the experience of not-going and of not-conducting time forward.

Chapter 30

Touching the dynamic of time by going deeply into the pain of not experiencing that dynamic.

Chapters 31–32

Drawing on the past, both positive and negative, as a source of knowledge, we can turn time toward inward knowledge. Inviting knowledge, exploring the future and the unknown, we can restore light to time and can join the lineage of the knowledge-holders.

Sacred Dimensions of Time and Space

Time 6

If time had 16 dimensions, how could we contact them? We could at least look for moments between moments, or time that departs from the timeline.

Time 7

Different ways of expanding or opening time, of finding layers within time and opening cones of time.

Level Two
Knowledge

Time, Space, and Knowledge

TSK 234–48

We can criticize and challenge the whole lower-order realm, without making any new claims, by seeing that "it is all just 'lower knowledge'." In this way we arrive at a knowing that is not the property of an isolated knower knowing specific things. Through such mindfulness or appreciation we can escape the limiting 'cones' of conventional knowledge.

TSK 255–60; Ex. 27–31

Challenging the usual structures, we can open the restrictions that seem to block us off from knowledge: in particular, the notion of a knowing self and of not-knowing.

TSK 271–85; Ex. 32–35

Through exercises, we can discover knowingness, clarity, and light as primary: not conveyed by things at all. We can cultivate this clarity to deactivate mind and self, drawing closer to a different kind of totality.

Light of Knowledge

Chapters 26–30

Knowledge could be different, because time and space could be different. But this requires challenging our commitment to the self and its world. We can see fear and confusion as pointers toward knowledge and as ways of challenging the logos. The move to the divine heads in this direction, but in the end cuts knowledge off. The mystical vision likewise leaves limits intact. We must aim for a more direct transition from level one to level three.

Knowledge of Time and Space

Chapters 13–15

Second-level knowledge can know the logos; it can know the non-specified rhythms of time, which go beyond the claims of substance and the measuring out of points.

Chapters 35–36

Second-level knowledge sees first-level knowledge bound up with first-level space, generating a temporal order that declares knowledge unavailable in space. This opens the possibility of dwelling in second-level 'zero space', before distinctions.

Chapters 37–39

Bystander knowledge emerges as part of a field given by second-level space. A knowledge that looks toward the allowing of space and the borders of field exhibitions need not be bound by the field; it can investigate field mechanics.

Chapters 57–60

Not-knowing could represent the potential for deeper knowing inherent in the knowing that shapes the logos of a particular field or read-out. It could be the carrier of knowledge, the 'x' of the axiom, disclosing knowledgeability. Zero stands as a symbol of this. Within each point is the zero point, the decisive point, where knowledge is about to take form: the point that grants existence permission to be. We turn from this point to follow out our concerns, and this seems inevitable. But concern itself can be a gateway: the gap in knowledge can become the root of knowledge. This leads to Great Knowledge.

Chapters 61–68

A critique of first-level knowledge that discovers the second-level knowledge inherent in the limiting and confining first-level order can become the expression of third-level knowledge. The key is finding the patterns inherent in experience and recognizing them as manifestations of knowledge; in turning from the self as bystander and allowing experience itself to manifest knowledge.

Chapters 77–80

We may have access to a knowing before name and form, available in our present minding, and especially available in times of transition such as the present era. We can consider it a 'zero level' of knowledge. We are most likely to find that knowledge in looking at the arising of name and form. Here we discover a giving that could also be given differently, a giving that does not establish what is given.

Chapters 81–84

Looking at right angles to labels, we see how knowledge arises within specific fields, such as the fields of the senses or the field of feelings. Fields have their momentums, but we can challenge their patterning by looking at the transitions between fields. Going beneath the content of the field, we arrive at time, space, and knowledge as they manifest in the field. We can arrive at the field of each field, turning from the content of the point to the decisive point of creation.

Chapters 85–86

We can trace how feedback leads to the projection of positions; we can see how gravity emerges out of zero gravity. We can observe how the subtle awareness of the senses emerges out of a wanting to know, giving a reality confirmed by symbols and language. But that is only the first stage. The senses can deepen toward Great Knowledge.

Chapter 90

We can see how first-level realities are built up by second-level mechanisms [a fine summary]. Touching the feel of the field leads us to the creativity of the mind, opening into time and space.

Chapters 92–94

Conventional knowledge gives expression to the order given by the logos. We can know the logos through cognition, coming closer to its fullness. But we can also know through the logos, or with the logos, or embody the logos. The not-knowing that is the fruit of inquiry is such a knowing. An example: the interlacing relation of subject and object. We arrive at a knowledge that embodies the knowingness inherent in the situation as a whole.

Visions of Knowledge

Introduction

We must learn to know without relying on concepts; to inquire in ways that evoke imagination and creativity; to use language differently and challenge all reference points.

Chapters 1–9

We need to proceed precisely, engaging heart and head, working with time, space, and knowledge themselves, inventing a new mind. Instead of shutting out not-knowing, we can rely on it. Logic says there is no escape, but gaps and transitions release us from this trap. We can touch the texture of space, experience the dynamic of time, and see in the light of knowledge; we can learn to improvise on what reason offers, to draw on thinking and turn to time and space as we embody them. In a time-space-knowledge telling, no restriction binds us. We can activate the vastness of the mind within its ordinary operations; can rest in the open. Ignorance is not the enemy.

Dynamics of Time and Space

Chapter 3

The field communiqué establishes only itself. Appearance and its structures lay claim to form, but take on substance only when we accept the story being communicated. The true liveliness, the source of freedom, lies in the telling.

Chapter 5

We can imagine space projecting images inward to the mind, and imagine mind projecting outward, without losing contact with space inwardness. Projecting the inward-outward of eknosis, we allow for space knowledge.

Chapter 6

Flattening space, so that each point is zeroless, gives appearance without substance. The claims of substance and of the knower can no longer be sustained, and the split between the 'one' of substance and the 'zero' of subjective knowing has never happened. Universal knowledgeability operates.

Chapter 16

Applied to the mechanisms of time, eknosis reveals time as dimensionless, opening the possibility that we can inhabit the inwardness of time.

Chapter 17

When we set aside judgments and give up the burden of being knowledge-holder, an intrinsic knowledgeability is available. Turning inward toward time and space, we connect with ourselves, not relying on precision but on the deepening of what is indeterminate.

Chapter 20

The path to vastness light opens through sinking beneath what appears, fearlessly conducting each new position.

Chapters 25–26

We can open the thoughts and structures of thinking and senses to knowledge.

Sacred Dimensions of Time and Space

Space 5

Second-level knowledge recognizes the ratio as the perfect manifestation of space.

Knowledge 4

To break free of the restrictions of the cone, we could explore how the subjective cone of knowledge links to the global time-space-knowledge cone. We could also consider the possibility of not being bound by the cone at all, just as time is not bound by the cone of history when we go the prior.

Knowledge 5

Second-level knowledge corresponds to second-level space, to which zero gives access. It is right knowledge, knowing the whole. Based on the mathematics of zero and cone, it strips the 'is' of its gravity. Every point offers this possibility, via its 'axes of access'.

Knowledge 6

16 is not a conventional structure. Available in every point as the fullness of zero, it grants knowledge access and offers the ratio, making possible the clarity of the decisive point.

Knowledge 7

The first-level ratio maintains the whole, but does not know it. The transition comes in recognizing knowledge as intrinsic to the order. We require a different geometry of knowledge, built up out of point and line, zero and 16, axis and ratio.

Level Three
Space

Time, Space, and Knowledge

TSK 13–18, 42–46

Great Space is the highest expression of the principle of openness that seems central to space. It is space with no concepts or distinctions, even though concepts and distinctions may manifest. All the familiar things constituting phenomenal appearance are Great Space *directly*, not by some remote or transitive link.

TSK 111–14

A review of how Great Space can be approached, at three different levels.

Love of Knowledge

Chapters 39–42

An analysis of space possibilities in terms of the insistence on substance and opposition, proceeding by way of negation and the 'ness'.

Knowledge of Time and Space

Chapters 19–21

In the zero-point of Great Time, space becomes the intimacy of time, opening into Being. From this openness, 'isness' presents itself to be known, but zero remains available: the dynamic of time excavating space.

Chapters 49–56

Great Space as complete allowing. When we follow the hollow, we arrive at nothing to say, and so enter the great allowing matrix. Tracing the lineage of appearance, we find that existence is responsible for its own appearance. We can inhabit the zero singularity of space, never establishing, only permitting. In fact, we cannot do otherwise, for we cannot escape the spaciousness of space.

Visions of Knowledge

Chapters 7–8

Where conventional space is devoid of knowledge, in the heart of space, the light of knowledge illuminates all. The body of space does not make itself available to be measured, but space remains 'free-occupied'; it is there in sense perception as the echo; it turns existence into metaphor.

Dynamics of Time and Space

Chapter 7

By abiding in thought without regard for the substance of thought's content, we can make thought into the agent of zeroless space, opening the prospect of freedom: appearance without substance.

Chapter 8

Zeroless space opens into knowledge. The finity of dimensions makes knowledge available everywhere, putting structure and content on the same level.

Chapter 28

Entering the stillness within appearance and learning to dwell in the Body of Knowledge.

Sacred Dimensions of Time and Space

Space 4

Zero as symbol founds 16: Symbol becomes 'fact', giving form while preserving zero.

Space 8

The specifics of 'space allowing' turn on the creativity of the zero-point, carrier of knowledgeability. Space is the 'may' to to time's 'can'. Opening to space, we open to knowledge and our own being and discover the finest space protection.

Level Three
Time

Time, Space, and Knowledge

TSK 98–101, 159–62

Great Time is the vitality of appearance, diffused and distributed out by a stream of significations and meanings. In Great Time, intimacy simply is, offering knowingness.

Knowledge of Time and Space

Chapters 19–21

In the zero-point of Great Time, space becomes the intimacy of time, opening into Being. 'Isness' presents itself to be known, but zero remains available: the dynamic of time excavating space.

Chapters 22–24

Second-level time reveals momentum, mechanisms, and the logos of the temporal order. Third-level time cuts through all this, transforming rhythm into the knowing of the Body of Knowledge, taking form in the mechanism of knowing. In the hidden dimension of time, past, present, and future merge.

Chapters 102–3

Inquiry touches the giving point of time as the Body of Time. Each emerging instant merges with the momentum of time. The rhythm of time is mastered, giving stable knowledge, invariable time, and space presentation. In the luminous patterns of the temporal order, points become symbols.

Chapter 108

The decisive point remains open because it comes into balance with the rhythm of time.

Dynamics of Time and Space

Chapter 14–16

A tempo transmits the temporal order—the base of the cone. But unknown time, outside the cone, transmits the tempo. To discover it, we must inhabit present time fully, replacing identity with the dance of time. "When the story is its telling, appearance is its appearing." (p. 147) We arrive at nuclear time and the lightness of being; we draw on imperfection to activate knowledgeability. Words fall short, but logos takes on new power. We can inhabit the inwardness of time, a nameless, creative arising, made available through the transformation of the experiencer into silent light.

Sacred Dimensions of Time and Space

Time 8

Time transitions through the power of zero. Perhaps we start over each moment; perhaps transition takes no time. A precise ratio connects 'from' and 'to', but all can be condensed back into the starting point. If space is space and time time, all points are connected. 16 supports the unfolding of time. Knowledge operates the ratios, and every transition is a new chance. Movement moves, but not from here to there. Points manifest as reflections. 16s share 16s; time shares space.

Level Three
Knowledge

Time, Space, and Knowledge

TSK 60–62; Ex. 13

As we discover experientially the openness of Great Space, Great Knowledge comes into view.

TSK 211–21

Great Knowledge knows only Space and Time, but leaves nothing unknown. When Space and Time are directly appreciated, that is automatically knowledge.

TSK 251–55

Great Knowledge does not do; it simply is. In the Body of Knowledge, nothing needs to change, and no self is required as knower.

TSK 286

Great Knowledge gives the uncontrived intimacy of the Body of Knowledge and an ongoing self-transcendence, with the flavor of ecstasy. We arrive at Being, the unity of Space-Time-Knowledge. A summary formulation: One point is all points.

Light of Knowledge

Chapter 30

Bypassing knowledge alternatives, we can arrive at level three by looking at the limits on level one. But we must admit we are caught in a web; that we *are* our limits. Our position (lower-level space) and our conditioning (lower-level time) close us in completely. Investigating these structures more fully, at a deeper level, we prepare for its erasure.

Chapters 33–38

In a more direct form of inquiry, none of the structures of conventional knowledge are limits, because each of them is available for investigation. The activity of knowing (cf. tracing) intrinsically embodies knowledge, and limits are the carriers of knowledge. [Although this analysis is preliminary, it already embodies level-three knowing.]

Chapters 39–50

Starting with negation, knowledge can steadily challenge every position. This is a level-three approach, even though it makes its way through level-two structures. It is a preliminary kind of practice, but it need never be set aside for something more advanced.

Knowledge of Time and Space

Chapters 19–21

In the zero-point of Great Time, space becomes the intimacy of time, opening into Being. From this openness, 'isness' presents itself to be known, but zero remains available: the dynamic of time excavating space.

Chapter 55

Knowledge that embodies Space fully becomes the Body of Knowledge. Appearance exhibits the quality of light,and knowledge appears as beauty.

Chapters 60–61

Great Knowledge manifests when we let the concern that makes us turn from second-level encompassing knowledge (the knowledge of not-knowing and the decisive point) become the gateway to knowledge. The gap in knowledge between knower and known, subject and object, becomes the root of knowledge. Great Knowledge comes through inquiry, but is not a goal: it is a gesture made by knowledge itself, manifesting in acceptance, balance, and appreciation.

Chapter 68

Great Knowledge comes with the turn away from the self and toward the interaction of time, space, and knowledge. We see knowledge as pervasive medium, and recognize the read-out as read-out.

Chapters 69–72

We can leave the open open, inquiring into mind and its interactions and into multiple dimensionality until we arrive at no measurement, nothing to point out, and nowhere to go.

Chapter 76

Negation as the opposite of affirming only assures that positions proliferate. But double negation does not affirm. It relies on *realization* to unite 'is' and 'is not', deepening awareness.

Chapter 80

Within the giving of name and form, we discover a giving intrinsic in the dynamic of each point, a giving that does not establish but grows out of the rhythms of knowledge and gives those rhythms.

Chapters 87–89

The senses are the agents of knowledge. They establish, but the original intimacy remains available. Whether we can open to this intimacy is a test case for Great Knowledge. We can trace this opening through nine stages, ending in the fulfillment of a perfect love. As agents of inquiry, the senses can go beyond all points to the Body of Knowledge. They can project the zero point of knowledge without projecting 'into'—project interaction and intimacy.

Chapter 91

Experience and awareness arising together give experience 'of', with no 'order' to be known.This time-space-knowledge interaction give the simplicity and radiance of 'only knowing'.

Chapters 95–97

Great Knowledge is not a result. It may develop gradually, but there is no procedure, no structure, no restriction. "The vision is in the seeing, and the seeing is in us."

Chapters 98–103

Every body of knowledge has its limits, but by releasing, we arrive at the limitless knowing of the Body of Knowledge. There is no taking form, no derivation. A lower-level sign of this is co-existence. In the intimacy of co-existing, "the 'non-existent impossible' can be created as unique." Nothing is held back. The Body of Knowledge offers the gift of light.

Chapters 104–106

Knowledge can be the background of what appears, the inner light illuminating all that is known. We need not choose, need not rely on beliefs.

Chapters 107–109

Knowledge traces the whole. It never closes off creation, which manifests in the decisive point: the gathering point of the whole and the ongoing emergency and the zero point that remains available even when all decisions have been made.

Chapters 110–112

Through inquiry we can enter the fullness of nothing at all. We can rely on inspiration and aspiration, and cultivate appreciation, compassion, and ultimately Great Love.

Visions of Knowledge

Conclusion

Knowledge is not just a tool or an instruction manual; it is a part of being. We can know as we breathe. Inquiry activates knowledge: "It is the essence of serenity, the heart of faith, the spirit of science, and the guts of practical achievement." We are always free to explore, to abandon commitment to our own concerns. In the freedom of no fear, we can inhabit the universal unique.

Dynamics of Time and Space

Chapter 8

Zeroless space opens into knowledge. The finity of dimensions makes knowledge available everywhere. With structure and content on the same level, we remain ahead of the possible.

Chapter 18

To challenge bystander knowing and kangaroo mind, we can challenge knowledge as arising in time. Knowledge as prior takes us back beyond the zero-point of origination to the zeroless. Nothing can limit such rainbow eknosis.

Chapter 19–20

Sheer speculation: light transmission as the sourceless source. Discovering this lightness, we discover beauty and joy. 'From' and 'to' fuse; we see light as the appearance of light. "The gods can only be jealous."

Chapter 21–24

We are free to enact light transformation, accepting each appearance as lightful. We transcend transcendence, and find there is nothing to do. Each limit calls out for transcendence, each point opens into vastness and intimacy. We touch the best we can.

Chapter 28

Entering the stillness within appearance and learning to dwell in the Body of Knowledge.

Chapter 33

Within the manifestations of time, the unknown of the future, the darkness of the unknown, we can invite light and find knowledge everywhere.

Sacred Dimensions of Time and Space

Knowledge 3

If we open the cone of knowledge, the point of origin and baseline disappear, and time and space become everywhere available. Knowledge embodies their dynamic, revealing an order we do not impose and overcoming the limits on knowledge.

Knowledge 8

When ratios fall away, we move toward a knowledge of knowledge that includes 'I am here'. The point of origin returns to zero, and transition becomes 'phase transition'. We cannot *enter* this dimension of all possibilities—the qualities that would grant entrance are not our qualities. Yet we stand at the juncture of time, space, and knowledge, and all is available. If we engage without a ratio of subject to object, we can see the mechanisms in operation. Ratio can merge with no ratio.

There are three keys: All points are symbols of zero; we can experience the depth of being; and knowledge unites with the expanse of experience. In this third-level transformation, 'I am here' enters the depth of fullness.

Three Levels in Sixteen Aspects

The sixteen charts that follow are based on the sixteen chapters of the first TSK book. For each chapter, key ideas related to the three-level structure of the TSK vision are identified in the six books that present the TSK vision. In most cases these ideas are given form through quotations from the texts.

The charts have entries for all six books, though not every book is referenced for each chapter. For each of the books, the key ideas are presented in terms of first-level, second-level, and third-level insights, with each level assigned a different color. For TSK, ideas relating to all three levels are shown on each chart. For the other books, from one to three levels may be shown. Related readings and suggested exercises are listed in the column at the right. For KTS, references to exercises refer to the chapter in which the exercise may be found.

Deciding whether a particular idea relates to level one, level two, or level three can be somewhat arbitrary, for almost every idea in the TSK vision operates on several levels at once. Still, this approach may help you trace the way the presentation of the vision unfolds in TSK and across all the books.

One interesting element in these charts is the way the quotations serve to evoke the vision. Each quotation is an invitation to let inquiry unfold in a particular direction. Of course, it would have been possible to choose countless other quotations, each one generating a different inquiry. The charts presented here are only one possibility.

TSK	"Whatever happens, wherever we go, whatever we do, there is the sky."(3) Space as always available, always allowing, and always found both within and beyond all seemingly substantial entities. In the end, that is all: "Space is projecting space into space." (10)	Ex. 1–3 (see also p. 24)
	The objects we perceive occupy a particular space, based on the "focal setting" in effect. Objects and their space are given together. As long as we focus only on objects, and not the particular space they occupy, we can never get to the bottom of existence. Cf. read-outs; "given together."	Ex. 1–6
	Our knowledge and appreciation correspond to the space available. We live in lower space, shaped by dualistic concepts of "something" and "nothing" (e.g., objects and space). But by moving through different spaces, we can ultimately activate the vision of Great Space.	Ex. 3–10 Ex. 26 cf. pp. 42–46
LOK	"We accept known limitations in space and time as final, establishing a focal setting and accepting a read-out." (414)	re: 'lower knowledge'
	"The rhythms of time and the allowing capacity of space respond to the quality of knowing in operation." (388)	Ex. 48A cf. Ch. 41
KTS	"Ordinary understanding turns away from the vastness of the 'not' toward positive descriptions of the objects it identifies." (123) "When the features of the known world are not, this is space." (129). Space and form as structurings of space. (130)	Ch. 25–28
	Compare "field" (also in DTS)	Ch. 37–44, 83–4
	"Within space there can be no gaps. What would they consist of?" (227)	Ch. 29–30, 49–56
VOK	"When we analyze what the senses give access to, we are reading the echo of unknown space."	Ch. 8.3 (153)
	"If space had qualities, it would be a specific 'something', . . . incapable of allowing." (15)	Ques. 1–25
DTS	"The training we receive as human beings teaches us to live in accord with the dictates of substance. We see with 'form eyes' and move with 'form bodies'." (29)	Ch 4–6 Ex. 7
SDTS	"In love with space, we are surrounded by warmth and wonder that cannot be lost, only forgotten. Space is our pervasive partner, closer to our being than any form. In the warm openness of space, learned meanings and imposed purposes give way to spontaneous nourishment." (63)	Ch. 1:8
	Compare ratio and cone.	Ch. 1: 4–5

TSK 1 The Presence of Space—Openness and Solid Surfaces

TSK	"As an alternative to [the usual view of the body as made of insentient matter, knowable by a self], it is possible that some kind of 'knowing' is the *primary* fact of embodiment." (24) We can cultivate this knowing through a focus on interactions and translucence. (27; cf. 44)	Ex 3–9 Ex. 11–14 Ex. 26
	Defining boundaries rob appearance of vitality and of knowingness. (29) In a more open view, there is no before, no pattern, no freezing. There are only tendencies to consolidation, marked by the emergence of the person who knows. (33-34)	Ex. 1–6
	The self that observes, experiences and "knows" results from "coming out" of a completely open dimension. (37-38) Physical embodiment likewise represents a never finalized consolidation tendency of the openness of Great Space, informed by Great Knowledge. (40–43)	Ex. 8–10 Ex. 17 cf. pp. 61–72
LOK	"The limits on the self's knowledge confirm the self's active control over knowledge." (251)	Ch. 30–31
	"'I' find myself reflected everywhere, for each known object . . . is 'my' creation." (195)	Ch. 24; Ex. 3–5
KTS	"The knowing that knows separation is itself a projection of 'space'. The act of attributing qualities to what appears is an aspect of what appears." (169)	Ch. 33–35
	"When the structures of meaning and existence become newly 'spacious', what becomes of the restrictions on human being?" (129)	Ch. 25–27 Ex. 28, 52
	"The self shapes the rhythms of time toward establishing its own existence." (26)	Ch. 4-6; 65–68
VOK	"When we look into the operation [of time, space, and knowledge] we are . . . exploring our own neighborhood. Our concern is with our humanhood, and our discovery is that spacehood, timehood, and knowledgehood may lead directly toward liberationhood." (98–99)	Ch. 3.3
	"Since every limit is referred back to the self, the self chooses limits by its very nature." (141)	pp. 141–43
DTS	"What appears can . . . be understood as a . . . 'being-reflected back' from space into mind." (39)	Ch 6; Ex. 1, 20
	"By choosing to 'point out' what is happening, we [frame] a reality that has no heart." (24)	Ch. 3.6
	"In successive acts of recording, what has come before is reprinted, preserved, and duplicated." (119)	Ch. 13; Ex. 11–12
SDTS	"Suppose there were a different space, a 'beneath' of physical, first-level space . . . Could zero . . . be a transition to that space?" (185)	Ch. 3:5
	"The whole construction based on the 'I am here' . . . remains fully reducible to zero." (5)	Intro., Ch. 1:1

TSK 2 The Body and Human Embodiment

TSK	The source of experience is not some item *within* the standard world view such as mind or self. There is no 'source from', no linear causality. Experience itself should be seen as proving this *conclusively*. (54) Singling out self as isolated knower only perpetuates this lower-space view.	Ex.11
	Self or mind purports to be the knower, taking a position vis-à-vis a passive external space. But space is better viewed as active, "a concomitant feature of all experienced happening." (50) Mind "is a sign of a particular focal setting being taken on Great Space." Experiental exercises develop sensitivity to 'setting being in operation' by working with its outputs. (64)	Ex. 11–14
	Thoughts need not do the knowing or meaning-giving. Great Knowledge is 'here'. Mind simply accommodates: "Mind, like space, has no foundation." (64)	Ex. 12–14 Ex. 1–9
KTS	"While time may be the active agent, the self that 'minds' need not be the one acted on. If it is time that reports experience, there need not be a subject of the report, or even an owner." (34)	Ch. 7, 33–35 Ex. 36
	"If space allows the appearance of the self as distinguished from its body, does this mean that space itself is alive or even intelligent?" (143)	Ch. 30 Ex. 56
	"Just as an exhibit in a museum exists both as exhibit and as the specific work it claims to be, so Great Space 'exhibits' appearance both as an exhibition by 'space' and as what it 'is'. (231)	Ch. 49–51 Ex. 52
VOK	"There is no need to insist on order and thus presuppose disorder; no need to strive for certainty and thus invite confusion." (132)	Ch. 6–8
	"Mind must enter an open trance. . . . Trance crosses over to transformation. It is the positive, healing response to the uncontrollability of ordinary mind . . . In trance we transmute." (147)	Ch. 8.1–2
DTS	"Could we shift our focus from the content of what arises to explore the movement and rhythms of the mind?" (xxxi)	pp. xxix–xxxv Ex. 4
	"Appearance manifests space: What appears 'in' space is space in operation." (32)	Ch. 4
	"Intent on . . . 'appearance as substance', mind affirms a 'source-from' understanding." (246)	Ch. 25, Ex. 1–3
SDTS	"We are reflecting on the origin of the origin, something that does not 'exist' in any ordinary terms . . . We might say that the origin of the origin is located in a fourth dimension." (27)	Ch.1:1–2
	"Zero stands for the holding power with which space endows appearance: the power to manifest space and form, existence, appearance, phenomena, and reality." (184)	Ch. 3:5

TSK 3 The Mind and the Origin of Appearance

TSK	"No source from" leads to "no 'by-standers' and no 'outside standers'." (72) "'Here' is a distracting by-stander in a world of 'outside-standers'." (74). The challenge: no first cause or causal sequence, no established world, and no spatio-temporal background that establishes 'here'.	Ex. 18, 19, 22A Ex. 24–26 Cf. Ch. 3
	Great Space is 'here', but not in the by-stander sense. Religious perspectives give some limited access to this 'mysterious' here by deactiving self-oriented focal settings, and thus contribute to knowledge. But all 'goal and progress' orientations" (80) must be deactivated.	Ex. 17, 18, 25, 26, 29
	To change the focal setting is self-defeating, if *we* do it. But there is a way to let go of the focal setting that neither affirms its output nor replaces it with an intensified focus on 'the present moment' or discrete sense experience (outside-standers). The 'given together' insight. (86–87)	Ex. 14, 15 Ex. 23–26
LOK	"Since what is 'real' is 'only what exists in the present moment', does this mean that being is recreated in each moment? . . . And if this is so, what has become of . . . cause and effect?" (182)	Ch. 22
	"If the 'I' results from a 'knowing' that sees and projects, it forms a part of the 'read-out'. How can it possibly bind us?" (360)	Ch. 28–29, 44 Ex. 34, 42
KTS	"How can it be established in advance that assumed regularities in operation will reliably come into play? From the first level, no answer is forthcoming, but in terms of 'field allowing' we might say that the 'field dynamic' assures these regularities *by not preventing them.*" (201)	Ch. 41–43 Ex. 44
	"The concerns and predispositions of the knowing subject—the 'bystander'— . . . reflect the interplay of underlying 'dynamics' and 'mechanics'." (195)	Ch. 41, 42 LOK Ex. 16, 25
DTS	"If knowledge is prior, the 'source from' model does not apply in the same way. We should be able to go beyond the zero-point of origination to a knowledge that is zeroless." (189)	Ex. 18–19
	"Like echoes bouncing off the walls of a cavern, appearances interact into one another, creating experience as a constantly shifting composite . . . These co-referring manifestations together make up the reality of the whole." (16)	Ch. 2–3 Ex. 5–6
SDTS	"We might try visualizing a cone of knowledge in operation in each single act of knowing, . . . [becoming] aware of the subject together with the object. Perhaps this single change would already be enough to make time available somewhat differently." (154)	Ch. 3:1
	"Ratio seems connected to the implicit structuring that pervades the world, including our perception of that world. Without [it], our experience would be like thick pea soup." (39)	Ch. 1:3, 3:6

TSK 4 The World and Other Bases

TSK	Challenging partitions and presuppositions is not a preliminary: It can be the path to Great Space. (91–93) This approach facilitates no grasping at the content of what is 'here', and then the infinite energy of a different 'here' becomes available.	Ex. 1–6, 11–12 Ex. 20, 30 Ex. 15
	Great Time is the dynamism that the openness of Great Space allows. It is not bound by the one-thing-at-a-time, from/to view of lower knowledge, which depends on meanings (102–4), nor is it bound by the dichotomies and isolationism of existence/non-existence.	Ex. 16
	Space, level one, "is precisely our ordinary realm, seen in the light of further possibilities for human fulfillment." (111) It sees appearance as an output of Great Space. Space, level two, is the experience of this link. Space, level three, finds no 'lower' space.	Ex. 16
LOK	"The self occupies a *place* 'here' and lacks knowledge regarding something *located* 'there'. This lack sets up a basic tension, which generates a momentum 'outward'." (263)	Ch. 30–34
	"If knowing takes place within a momentary sequencing, each passing moment . . . its knowing capacity and energy, [are already] 'used up'. The dynamic of the present slips away. (189)	Ch. 21–23, 42 Ex. 23
	"'This' is not 'it'; . . . there is no 'it' and no holding onto something as established." (366)	Ch. 45
KTS	"If the 'emptiness' of 'space' is *inseparable* from the meanings we assign existence and [from] 'objective reality' . . . , do meanings, identities, and values still play the same role?" (129)	Ch. 25–28 Ex. 52
	"We are looking for explanations, when knowledge lies elsewhere . . . Mind *gives* experience."	Ch. 90
	"If, in the course of this inquiry, the old assertions prove hollow, we can follow the 'hollow' into the allowing negation of Great Space." (229)	Ch. 49–51
DTS	"Suppose we could flatten space, eliminating its dimensionality . . . All physical phenomena are left zeroless. . . . A zeroless world [offers] neither blockage, restrictions, nor borders." (42)	Ch. 3, 6 Ex. 7
	"Each story we tell is also a story about the way that time functions . . . The conductor conducts into being a time that supports the telling of stories, and then conducts [accordingly]." (125–26)	Ch. 13 Ex. 11–12
	"In the uninterruptible openness of zeroless space, [we can] stay a step ahead of the possible. Instead of depending on, we can move out into." (67–68)	Ch. 8 Ex. 7–10
SDTS	"'I am here' does not confine us. How can it, when it is simply the front side of zero, the mask through whose mouth we first hear of the point of origin?" (171)	Ch. 3:2–3

TSK 5 Being in the World, Being Space and Time

TSK	The law-like regularities acknowledged by modern thought can be seen as expressions of the structure of lower-level time, which presupposes sequence as an inexplicable given. Time simply passes. Typically we see it only as the index for measuring what happens.	Ex. 18–22 cf. p. 81
	Lived time is 'lower time': "the appearing and transforming of the . . . components that constitute a situation," (124) locally partitioned and arranged into an inexorable past-present-future sequence, in which the present is given priority. The outcome is helplessness and frustration.	Ex. 7–10, 14 Ex. 18–22
	Great Time is ever available, for it gives all components of each situation, including the transitory present moment and the final partition of death. Time as index reflects one aspect of such giving: Things *are* such measured-out series, with which we are "timing ourselves away." (127)	Ex. 19 Ex. 25–28
LOK	"If the linear structure of time is open to question, the basis for the world known through descriptive knowledge collapses." (125)	Ch. 15 Ex. 13–16
	"Whereas space is central to the world of objects, time is central to the world of the self." (150)	Ch. 18, 23
KTS	"Linear time is the 'carrier' for experience, assuring through its specific 'carrying capacity' that experience will be experienced in a particular way." (32)	Ch. 2, 7–9
	"Perhaps the future gives the present form . . . while the past gives it substance." (9)	Ch. 1–3
	"Existence itself is an *activity* . . . a gathering momentum, a dynamic being. The classification into past, present, and future is a very rough approximation of this fluid motion." (45)	Ch. 16–18
VOK	"Interpretation unfolds retrospectively; we cannot say what is happening in this instant." (121)	Ch. 5.2
DTS	"The music is already playing, [and we must conduct accordingly]. Bound by [the past], we can act creatively only within the shadow zone at the borders of the recognized." (118)	Ch. 13 Ex. 17–19
	"In the moment we specify the present, we turn it into the past . . . All of time is past time." (84)	Ch. 9–10
	"The indeterminate interplay of mental constructs [presents] the openness of the future." (130)	Ex. 13–14
SDTS	"Because we do not know the motions, the rhythms, or the transitions of time, we have never solved age-old questions about time's creative mechanisms." (76)	Ch. 2:2–3
	"Following time as a one-way stream, we see only one possible cone of development . . . But if knowledge knows how to operate the ratios, every transition is a new chance." (134)	Ch. 2:8

TSK 6 The Presence of Time—Liberating Potency and Compelling Patterns

TSK	Time, Stage One—'lower time' as abstract index and compelling force—"amounts to an attempted taming and appropriation of Great Time for egoistic purposes." (137)	Ex. 7–10
	Time, Stage Two abandons meanings for experiential acquaintance with time, and discovers it as the source of inspiration and spontaneity. An initial sense of time flowing gives way to a focus on presence that reveals the hidden dynamics of 'timing' (145). This discovery allows for discontinuities and singularities, via "feeding time into time." (149)	Ex. 22–23 cf. Ch. 8
	When timing does not insist on the link to sequential flow or 'going' (151), all points become accessible (155), and infinite possibilities open, based on intense intimacy. When all sense of happening and 'thing' gives way, this is Stage Three: "Intimacy is." (159)	Ex. 23
LOK	"The force of the self's desire unfolds as the momentum of linear time." (149)	Ch. 18–25
	"'Random' time is unitary, for partitioning is based on order." (243)	Ch. 29
KTS	"The self is 'of' time, 'at' time, and 'on' time. Each specific point in time, each 'now', is 'measured out', and each measure of time has its own reference." (38)	Ch. 3–9
	"The whole of appearance can be recast as simply this: *An exhibition rhythmically transforms* . . . In the intimacy of the instant, Space and Time open to Being." (86, 94)	Ch. 18–20
VOK	"Time is remote or overwhelming because we do not know how to join in its . . . dance." (88)	Ch. 2.3
DTS	"Let us imagine substituting for conducted identities and stories the image of appearance danced by time . . . The distinctions of the dance transform appearance. . . Identity gives way to the universal unique, expanding and transcending in a rhythm of finity." (146)	Ch. 15 Ex. 14–16
	"The Body of Knowledge manifests in experience as awareness free from the fixed momentum and established derivations that give rise to thoughts and images." (229)	Ch. 23
SDTS	"My 'am' is that space where knowledge and time intersect." (111)	Ch. 2:5
	"Why should time be compressed into a single dimension that belongs nowhere? Why are we compelled to travel in a tube . . . , instead of being allowed to roam freely in time?" (113)	Ch. 2:6
	"Time is everywhere time, not bound by here to there, past, present, or future. Not going anywhere, it is pervasive and penetrating. Activating existence in all directions, uniting everything in all reflections, . . . time leaves nothing static. Being is alive with time." (137)	Ch. 2:8

TSK 7 The Path from Ordinary to Great Time—An Overview

TSK	"The self is precisely the *embodiment* of a lapse of . . . understanding [Great Space and Time]." (166) When the self declares itself knower and thinker, "the power of time fragments, becoming the volition of the self . . . set in contrast to the energy of nature and physical processes." (169) What is dynamically 'timed out' (subject, object, distance-between) becomes a horizontal *index*.	Ex. 17
	The self is an impoverished version of the present, always doing and going. Tripartite time must be challenged by rediscovering the timing that gives the self. Then "points" of time disappear (cf. the zigzag diagram [179 ff.]) We find there is no 'one standard time-line'.	Ex. 18–22
	We *ignore* the open possibilities of 'going'. Appreciation leads to 'knowingness' and then to 'not-going': centerless abiding and intimacy with all times, directions, and 'things'.	Review Ex. 1–22; Ex. 23–25
LOK	"The self *establishes itself and the world that it knows* by moving off-center, 'beyond' the point-instant of temporal knowledge. In one unified action *it takes a position, posits a situation, and imposes meaning*. The self finds itself 'in the world'." (147)	Ch. 18–25 Ex. 1, 3, 5, 18, 19, 21, 24
	"If we focus on a particular interval between points A and A1, we find there is a moment of transition—'x'—that . . . remains unknown. Try investigating 'x'." (183)	Ex. 22–23
KTS	"The temporal priority of the objective realm is based on a prior accepting of the existential priority of the self." (40)	Ch. 7–9
	"Caught up in the idea of existent things—'outsiders' trapped in the rhythm of time—we make of movement a physical event, and miss the movement that is intrinsic and inherent." (93)	Ch. 19–20
DTS	"Our attempts to ignore time's passing are reflected with fidelity in the patterns that time presents. Things run down." (287)	Ch. 29 Ex. 11–12
	"As past, present, and future interact in nuclear time, they exchange places with one another, expressing the dynamic rhythm of interactions that remain non-dimensionalized." (156)	Ch. 16
SDTS	"The knowledge that seems most self-evident is this: 'I am here'. Suppose we make this knowledge our starting point. Can we let this fundamental axiom of first-level knowledge open into deeper, more penetrating knowledge?" (3)	Introduction
	"The ratio is the knowledge that time needs to move. And this movement is no simple 'from/to' two-step. It appears to be a complex dance in all directions . . . and depths 'at once'." (129)	Ch. 2:8

TSK 8 Intermediate Experiences of Time

TSK	Our ordinary encounters, reactions, etc. require us to ignore time's dynamic, making it a background against which identities and meanings emerge.	Ex. 17–25
	Each moment is potentially all-embracing, but in terms of content is unique. It is a 'read-out' that establishes nothing about other read-outs. A read-out is "the informing communicative output by 'time' of a particular 'focal setting' on 'space'." (195) No read-out or point establishes either itself or anything else. Read-outs are self-contained. Sameness is only a convincing message.	Ex. 25–26 Ex. 7
	If things are 'not-established', appearance can remain on the surface. (199) But this insight is itself only an "unusually incisive" read-out. *Approximating* Great Time, it initiates a process that sees through the 'read-out insight' as always transitional. Read-outs undermine themselves.	Ex. 7, 26
LOK	"[Through] the self's commitment to its identity . . . the founding story and the narratives it supports are maintained, but the dynamic inherent in those stories is lost." (205)	Ch. 25 Ex. 41–43
	"The conclusion that the 'read-out' is all-encompassing is itself a construct of the read-out." (316)	Ch. 38, 45
KTS	"Having set up a rational order, we make points and connect them, marking out the motions of the specific . . . The momentum of 'what is' controls our responses." (46–47)	Ch. 8–9
	"The read-out relationship . . . does not assure that conventional time captures anything fundamentally accurate about second-level time. In fact, the reverse is true." (66)	Ch. 14–15
	"It may be that Time is a total energy, invariantly embracing the whole, but what will allow 'us' to embrace this possibility?" (110)	Ch. 22–23
VOK	"Feeding on its own energy, reflecting its own conceptual structures, mind is left to make sense of its own images." (106)	Ch. 4.1
DTS	"Events reflect images into thoughts, and the transmission continues." (87)	Ch. 10, 13
	"Relying on conventional 'biofeedback', we enter into the pronounced projected." (150)	Ch. 15
	"If time is truly in some sense source, it is immediately present and available. As the mind goes about remembering the past and conducting the future forward, consciousness must somehow remain connected to the hidden time of founding." (137)	Ch. 14 Ex. 13–16
SDTS	"Imagine each point . . . fully endowed, completely initiated, and empowered to play out the fundamental ratios of existence." (132)	Ch. 2:8

TSK 9 A Self-Challenging Vision Opening to Great Time

TSK	Knowledge is both goal and basis of the path; it is "the immediate and knowing dimension of all reality and experience . . . Everything is directly Space and Time, and Knowledge is inseparable from these." (211–12)	Ex. 25–26
	"Knowledge is not something which knows something; it is simply the presence of reality as 'knowingness'." (212) Our beliefs and misapprehensions reflect this knowingness, but darkly, piling meanings on meanings till we feel cut off from knowledge and fulfillment.	Ex. 27–29
	When every ordinary presentation offers knowingness, the self can grow, and we can become more effective. There is no user or using of knowledge and no linear path to follow. "When Space and Time are directly appreciated, that is automatically Knowledge." (216)	Ex. 30–35
KTS	"Inquiry as Great Knowledge allows the Body of Knowledge to become the Body of Time, gaining intimacy with Time by embracing it in the instant." (457)	Ch. 102
	"In the knowing of 'knowingness', Great Knowledge presents itself." (443)	Ch. 98–99
	"When contradiction as a barrier to knowledge disappears, this is embodiment." (280)	Ch. 60–61
VOK	"Ordinary mind is deeply enmeshed in its own constructs . . . A radical shift is called for—a translation from one way of being to another . . . The mind must enter an open trance." (147)	Ch. 7.1
	"We need not put everything in its place. We can act on the intention to hold the unnamed open, so that limits do not become final." (143)	Ch. 7.2
DTS	"Knowledge opens knowledge into universal depth. Wherever we go, whatever move we make, we advance toward knowledge. Knowledge . . . embodies itself, drawing strength from the momentum of its own opening . . . It is the situation of 'nothing that needs doing'." (219–20)	Ch. 21
	"Imagination can be the point of entry, releasing the dynamic trapped within the temporal order . . . Once the time of our being gives permission, space knowledge proves available." (241–42)	Ch. 24
	"Through the union of time and space, knowledge can be born within human conditions, transforming their operation. *Being itself can become the model for knowledge.*" (241)	Ex. 23–24
SDTS	"The prospect for singularity comes from knowledge. In knowledge, each point is newly born . . . Time and space are available in the fullness of 16, poised before locatedness and within happening. The ratio has not yet been formulated. The transition endures." (223)	Ch. 3:8 Ex. (pp. 120–21)
	"The ongoing crisis can be our thesis." (220)	Ex.: pp. 219–20

TSK 10 The Possibility of Knowledge as Commanding Insight

TSK	Knowingness has the quality of perfection, "because nothing more . . . needs to be known." (219) When we try to *achieve* knowledge, we only end up freezing knowingness. "Until we relate to this dimension of reality, our actions and decisions will be clumsy and uncertain." (220)	Ex. 23–25 Ex. 29–30
	The critique of conventional models amounts to a blueprint for 'acquiring' Great Knowledge. Such critique is more than theory, in part because it questions the basis for theoretical knowledge. It exposes presuppositions as variations on a theme.	Ex. 4–11 Ex. 16
	Provisionally, it is useful to emphasize that all conventional knowledge is subjective, based only on 'mindings'. This helps reveal the dynamic configuration of lower time, its pervasive dissatisfactions, and the limits of proposed solutions. But mindings as basis can be challenged.	Ex. 24–26
LOK	"As analysis tunes in to the 'momentum of arising', starting from the point at which 'we' find ourselves and *working backward, it naturally reverses* the tendency toward setting up rigid positions . . . and the open, allowing nature of experience comes to the fore." (276)	Ch. 33 Ex. 1–2 Ex. 33–35
	"When such questions suddenly take on real significance, it is like discovering that we were adopted as children and raised by strangers: The familiar at once seems profoundly foreign. We face an ignorance so deep it is actually embarrassing . . . But . . . our embarrassment and not-knowing are what make the old structures accessible to inquiry." (282)	Ch. 34
	"We may recognize that even the complex of energy and labeling that bears the identity 'I am' is only another name, a patterned dynamic put into operation by knowledge." (294)	Ch. 35 Ex. 5
KTS	"One . . . indication [of Great Knowledge in operation] is loving acceptance—not only on the interpersonal and psychological levels, but in terms of the fundamental structures through which the self proclaims its being." (278)	Ch. 61–64
	"A nonpositioned, unbiased questioning reveals reality as . . . recorded determinations and distinctions arising through the transitional operation of indeterminate 'fields of knowledge'." (371)	Ch. 81–84
VOK	"Not-knowing arises not only at the boundaries we depend on for determination, but within the perceptual process that establishes the known." (93)	Ch 3.1–2
DTS	"In our present and extended thinking 'through', the conviction emerges that we are free to do differently, to enact differentness. The knowledge required is available." (239)	Ch. 24 Ex. 20
SDTS	"If [knowledge] commits to the way things are as the defining limit of its capacity . . . the inherent fullness of the zero-point will become the structured certainty of the starting point." (179)	Ch. 3:4

TSK 11 The Obscuring Nature of Ordinary Knowledge

TSK	The 'minding and belief structures' approach proliferates, obscuring all alternatives. But it collapses under the challenge that all we know is "all just 'lower knowledge'" (235): not 'what the subject knows', but the very trend that involves a subject knowing objects.	Ex. 22–26
	Our realm just is this 'lower knowledge', a constrictive take on space, time, and knowledge that cuts 'things' off from reality and makes them fictions. It is a meaning or belief, but not in the lower-knowledge sense. We remain bound to the lower-level realm by fear.	Ex. 25–31
	The critique of 'lower knowledge' opens its *knowledge* dimension to higher knowledge, leading to "appreciative intimacy" (241) as primary. Mindfulness (knowingness) of Space and Time and our steadily changing relationship to them work the shift. We escape the cone. (246–48)	Ex. 25–32
LOK	"Because the force of desire and wanting thrusts the self ahead into what might be, the possibility of enjoying what is already present is blocked." (155)	Ch. 13–20 Ex. 45–46
	"Confusion may be an ever present possibility: a cognitive emotional response to the unspoken and unthought recognition that beneath our stories lies 'something else'." (225)	Ch. 27
	"The magic that makes this transformation possible is not imported from the realm of the esoteric . . . It . . . finds its power in embracing both conventional knowledge and a knowledge that goes 'beyond' the conventional. Another name for it is the love of knowledge." (372–73)	Ch. 46
KTS	"'Not' can be understood as the fundamental tool of logic, which uses it to move in a linear way from point to point." (341)	Ch. 73–74
	"We do not need to say 'yes' or 'no'. We can 'realize' the object or feeling together with what it is not, letting 'is' and 'is not' arise as a whole." (349)	Ch. 75–76
VOK	"Within the idea there is opening the idea; within interpretation, there is the luminous interpretive act. Opening within time and space, awareness does not stay within the boundaries of what presents itself, but also goes nowhere else." (144)	Ch. 7.3
DTS	"[Knowledge] offers the dance and drama of appearance, the going and the return, the misunderstanding and failure to communicate, even the possibility that we are completely lost." (192)	Ch. 18
SDTS	"Zero 'reads' space or expresses it, in such a way that higher-level knowledge remains available. Zero-based operations allow for an interpretation of the ratio that does not give existence the same weight or gravity, . . . so that appearance can express the fullness of knowledge." (190)	Ch. 3:5

TSK 12 Lower Knowledge—A Hindrance and an Opportunity

TSK	"All ordinary sensing and knowing can be Great Knowledge, and so can instances of confusion or of 'not-knowing'." (252) It is not a message from Time, for such a message would not understand time. It is unlearned learnedness: the Body of Knowledge.	Ex. 27–31
	The Body of Knowledge has a sharp ecstatic quality, available in all experiences. There is nothing that needs to be controlled; nothing to use or to bring about. It emerges naturally, as certain common experiences suggest.	Ex. 29–31 Ex. 32–34
	Knowing comes from a different and unknown level, one that does not require a knowing self. "In a world seen as an open horizon of 'knowingness', all presentations can be positive, clear, illuminating, and nourishing." (260)	Ex. 29–35
LOK	"The task of inquiry is to open this channel of communication between not-knowing and knowing—to create the opportunity for knowledge to speak." (379)	Ch. 47
	"To manifest the 'Body of Knowledge' is to go 'beyond' meanings to the texture of all experience . . . the intimacy of the immediate intuitive instant." (409)	Ex. 46–50
	"We need to ask where 'I-ness' comes from. If the 'I' results from a 'knowing' that sees and projects, it forms a part of the 'read-out'. How can it possibly bind us?" (360)	Ch. 44
KTS	"Since knowledge tends by nature to know, why should it be limited at all?" (260)	Ch. 57–58
	"The Body of Knowledge is not a position body, an ownership body, a category body. Embodied Knowledge is extension and continuation, relationships and distinctions and character. It is the creation of the 'field', the 'feel' of the 'field', and the creation from 'field' to 'field'." (445)	Ch. 98–100
	"In each act of positioned knowing, there is a 'point of decision', a vital 'zero point' 'before' knowing is 'accomplished'." (270)	Ch. 59–60
VOK	"Suppose we chose . . . to focus on the pre-existing 'body of knowledge' out of which the knowledge we discover emerges?" (157)	Ch. 9
DTS	"In the interaction of awareness with time, an intrinsic knowingness is preserved forward." (139)	Ch. 14
	"Knowledge is available . . . in the twisted, the perverse, and the misdirected." (143)	Ex. 1, 2, 23
SDTS	"When we know the ratio of a problem, there is no longer any problem—only the ratio." (208)	Ch. 3:7

TSK 13 The Emergence of Great Knowledge

TSK	As a 'drawing apart tendency' and ongoing consolidation, the self *must* feel dissatisfied with the experience it labels. Even positive experiences display an unbalanced patterning: three strata that generate conflict and uncertainty. And accepting labels has its own negative consequences.	Ex. 7–10, 16
	Within all such limited experiences, we can cultivate a fuller knowingness, a "being *in* the energies which we *are*." (268) Then nothing needs to change—no "therapy" is required. An example: tasting the nectar of love not distorted by the consolidating tendency toward craving.	Ex. 32–35
	Knowingness as primary gives clarity, through which higher knowing capacities may emerge.	Ex. 23, 32–35
LOK	"To define is to be confined within the definition's narrow range." (105)	Ch. 13
	"A position is the outcome of an *act* of positioning . . . Seen in this light, *positions are expressions of knowledge, rather than structures that limit it.* Instead of accepting the viewpoint of the 'bystander', which insists on its fixed positions . . . , we could see in positioning *the manifestation of a knowing that is not itself situated or specified.*" (272)	Ch. 33
KTS	"Bound to a specific mode of existing, the self is committed at the outset, indeed for all time, to a specific not-knowing that fosters an ever present existential disquiet." (310)	Ch. 65–67
	"To take this possibility beyond the realm of the hypothetical, we could seek out a different kind of 'feedback'." (315)	Ch. 68
DTS	"We do not need to speak for time and space, for if we know how to listen, time and space speak for themselves . . . We can let go of the knowledge that we possess and merge with a knowledge that transcends all positions." (178)	Ch. 17 Ex. 20
	"We have learned to think of knowledge as linked to distance: to standing back and judging cooly. Now we know that knowledge is something quite different. It is the love we feel for all appearance, the love that unites all appearance." (196)	Ch. 18
	"The interplay of identities in the foreground of experience obscures the light transmission. But beyond identity and what we can identify, there is a way to experience light more directly. The 'lightness' of sensing can open to the nature of light." (204)	Ch. 19 Ex. 21–22
SDTS	"The turning point comes when knowledge expands to include the 'I' as knower, the 'I am here' at work fabricating the framework. (219)	Ch. 38
	"Dwelling in the balance point, knowledge discovers the place for being fully alive." (235)	"Three Keys"

TSK 14 Great Knowledge—Its Value and Pervasive Quality

TSK	Clarity rather than 'mind' and 'things' can guide us. We attain this 'knowingness' by working back from objects and experiences to the clarity they bear. No experiences are separate: "The previous is not previous; the distant is not distant." (282). There was never darkness.	Ex. 32–33
	There are three levels of relating to knowingness: 1) 'lower knowledge'; 2) approaching and embracing totality, with 'things' no longer viewed as isolates, but still assigned meaning; 3) no fixed totality, and no individual attainment, but only Space and Time.	Ex. 29–35
	The unity of Great Knowledge is the Body of Knowledge, an "uncontrived intimacy." (287) We critique ordinary mind, open to knowingness through experience, but then see that there was never any separate mind (or things) to begin with. "Their happening is their non-happening."	Ex. 30–34
LOK	"Negation discloses a new dimension to knowledge—a bottom-level 'zero-ness' emerges within 'ness-ness', while 'I-ness' gives way to 'knowingness'." (365)	Ch. 45
	"As inquiry and investigation unfold the dynamics of observation, interpretation, and negation, three different kinds of knowledge are revealed. We can think of them as three brothers." (365)	Ch. 45
KTS	"As knowledge and time and space interact, three levels arise . . . However, this is a first-level description. From a second-level orientation we could say that at each level a 'read-out' prevails and a field unifies; in third-level terms we might say that throughout time and space the Body of Knowledge is variously enacted." (429)	Ch. 95–97
	"The present is not stagnant, not determined. There is no 'between' as a separate position, no 'one' that is not 'all', no 'apart from' in what is unique." (296)	Ch. 63–64
VOK	"Mind opens to knowledge as the petals of a flower unfurl from the bud into the blossom. Then there comes a second opening, a deepening lightness that opens into space, revealing ultimately the heart of space—a realm of ethereal beauty . . . Here appearance itself opens into lightness, and opening lightness reveals the body of knowledge." (137)	Ch. 7.1
DTS	"At the edge of our thinking, the quality of light still makes itself known; beyond perception of the object, the perception of projection continues to be available." (206)	Ch. 19 Ex. 21–22
	"What starts as an idea discloses its availability at the center of all that is." (339)	Ex. 25
SDTS	"Sheer wonder at all the ways experience can embody deepens into the final wonderment that knowledge can unite the depth of being and the expanse of appearance . . . [In] the being of here and now, [knowledge engages] the being of the point. The 'I am here' enters . . . fullness." (236)	Ch. 3:8

TSK 15 Knowledge and Totality

TSK	Space, Time, and Knowledge offer a triple-faceted vision of the unity of Being, but not in the sense that Being stands behind them. The presence of Being overrides 'is/is not' distinctions. Being is no 'thing' and also no 'no-thing', no hidden ultimate.	Ex. 34–35
	Once we let go of egoistic purposes and values, Being proves inseparable from human being: "sheer art, beautiful beyond all enduring." (297) All actions become ritual gestures and light-hearted play. Newly born second by second, we are bonded with others, given together by time.	Ex. 23, 26 Ex. 32–35
	Our unity is fundamental, for one point is all points. This is the true basis for equality and for perfect spontaneity. Self-seeking makes no sense. We know intrinsically how to act, and feel all-encompassing compassion. Personal territory drops away, a destructive fiction. We enact Being.	Ex. 34–35
LOK	"When negation pervades knowledge, knowledge pervades being, revealing being as more fundamental than either existence or non-existence . . . Since knowledge is not used by the 'self-asserting I' to set up a specified reality, the distinctions that operate to confine knowledge do not come into play." (364–65)	Ch. 45 Ex. 46–50
	"Perhaps we could trace a more global knowledge 'in action'. First, it would be free of assertions, for assertions stop the free flow of knowing. Second, it would be free of judgments, for judgments require positions. Third, it would be free of distinctions . . . From the perspective of conventional time and space, such a knowing would not be a knowing at all." (408)	Ch. 50 Ex. 50
KTS	"We might ask why Time, Space, and Knowledge show up at all. In a sense, the answer is simple . . . The universe, history, and human intelligence presuppose them." (434)	Ch. 95–97
	"The 'no journey' perspective is not . . . an excuse for deciding that 'things as they are are perfect.' The point is more straightforward . . . The vision is in the seeing, and the seeing is in us." (427)	pp. 427, 431 Ch. 95
DTS	"Knowledge the Old Father invites us to grow to full maturity, ready to benefit others." (240)	Ch. 24
	"In the shifting interaction [of past, present, and future, a] nuclear force radiates outward, and the trinity transmits forward. In the truth of this transmission through time and space, one point is all points, one time all times." (156–57)	Ex. 16–17
SDTS	"How could knowledge not respond [to space and time]? Holding the key and taking the initiative, knowledge [demonstrates] where to draw the lines and how to draw the benefits." (61)	p. 61
	"In every moment, every point of thought can open, and every point of being can manifest." (237)	p. 237

TSK 16 Knowledge, Being, and Human Being

Summaries of the Books

The language in which the TSK vision is presented is vital to the power of the vision. Just as the plot summary of a novel leaves out most of what makes the novel a work of art, so the summaries presented here inevitably leave out what makes the vision an evocation of being. They also risk distorting what is said.

Still, the TSK vision is so rich in ideas that these summaries may be helpful as a kind of road map to the vision. If you find yourself lost in the middle of reading a book, you can come here for guidance. If you are deciding which book to read, you may find here some indication of what direction and approach each book takes.

For each chapter or section summarized in what follows, the corresponding page numbers are given. This may help you gain quick access to chapters referenced in other parts of this book, particularly for KTS, where the chapters are numbered only in the table of contents. (The CD that comes with When It Rains *also offers several effective ways of navigating through the books.) For all the books but TSK, links are given for each chapter or section to particular facets and levels of the vision.*

Note that an outline of KTS, more detailed than the summary here, has been published in VOK.

Time, Space, and Knowledge:

A New Vision of Reality

Chapter 1 3–18

The Presence of Space—Openness and Solid Surfaces

Space is within all objects, including mind objects; the kind of space we discover in our experience depends on the focal setting in effect. What we perceive as objects are defined in contrast to space, so if we focus on understanding objects only, we cannot understand either appearance or existence. "If we become able to change the setting, fantastic new knowledge and appreciation of life can be gained." (4)

Our understanding of space is confining. A new understanding, based on new focal settings, would itself be a kind of space. "An understanding—or vision—can itself be the revealing basis of all reality." Such a vision is based on Great Space.

Objects need space to exist, and the *seeing* of objects also requires a kind of space. But the intellect stays busy with objects, while space itself is untrackable and ungraspable. By contemplating space, we may experience more relaxation and satisfaction. "Can space itself be the fundamental nature of existence that we seek elsewhere or simply ignore?"

The relationship between space and objects is "an interesting and suggestive aspect" of our reality as a whole. It suggests the possibility that our ordinary reality expresses a more unified reality, but has unfolded "so as to give great significance to dualistic categories." But information pertaining to the higher, more unified order cannot be lost. It can only be "cast in a form that admits of ambiguity." This ambiguity means that "even conventional dichotomies . . . preserve [symbolically] expressions of a higher order unity."

An example is space itself, which relaxes and allows us to open up, while constituting "an intimidating blankness" if we pursue it by itself. This is made more suggestive by the fact that we can find more space simply by changing our focal setting. Used "in a certain way" these considerations are not just poetic daydreams; they are "pleasing and accurate symbols of a reality that cannot be shut out."

If we track existence with a sense of such symbolic presentations, "it seems that we have no choice but to consider a kind of space as somehow fundamental." Leaving logical analyses aside, if we contemplate comings and goings, we find that "Space is Projecting Space into Space!" (10) From ordinary perspectives, this insight is unacceptable. But it does not aim to reduce ordinary reality to space as it is conventionally conceived (see also top of p. 17). Instead, we should take the ordinary as a symbol of the higher realm. Contemplating space "actually guides our awareness into new types of spaces, until we can reach a kind of space that is, in more than a symbolic way, the ground of everything."

In addition to space as we observe it, a more pervasive lower space is in operation, a function of a particular type of focal setting. This lower space is responsible for the character of all events and appearances. It has a structuring power, whose effect leads us to lose out on the more fulfilling and significant dimensions of space, as though we were being ruled by a petty official. But our particular lower space is not the only one. There are infinitely many spaces, each with its own subtly limiting dynamics. These can be meaningfully ordered in terms of being more encompassing, open, and inclusive. Breaking out of one space seems initially to require force. In this hierarchy, Great Space occupies a unique 'position', because there all models of hierarchy and greater or less accommodation break down.

As we travel through more open spaces, we can discover a space that involves no concepts at all, "since concepts are simply indications of the relative opacity and resistance of a particular space." If thoughts, concepts, etc., conceived of as limiting walls, can be rendered transparent, without thereby setting up new walls, this is like Great Space. The distinction into partitioning concepts that applies to lower spaces does not apply to Great Space. There can be surfaces and walls with no colliding or bouncing off. It is difficult to grasp this simply by following a verbal explanation. "Ordinary languages have features that reintroduce subtle distinctions even in phrasing denials of them." We need to drop the logic of alternation and to break down the discreteness of things in lower space. The scientific view of cause and condition can be helpful here, but it tends to reinforce constricting notions such as causality. "The knowledge, language, and images of one level—by themselves—can never entirely recapture the insight of a higher level."

Something must be done. Lower space crowds and constricts, introducing disharmony. Our lives do not flow; we are blocked and frustrated. We have to see how we are accepting limitations unnecessarily. Then we can stimulate integration and creativity.

Chapter 2 21–46
The Body and Human Embodiment

The Giant Body exercises offer a way to gain appreciation for the familiar and challenge preconceptions.

Exercise 1: The Giant Body. Visualize a giant body.

Exercise 2: Internal Details: The structures, systems, and subsystems of the body are not things. They are mainly space, which itself accommodates and is defined by interactions between other smaller structures and systems.

Exercise 3: The Microlevel (atomic and subatomic particles): Conscious experience itself is considered to be a function of insentient physical systems such as these, yet "it is possible that some kind of 'knowing' is the primary fact of embodiment. The 'known', 'unknown', 'unknowing', and (directly) 'unknowable' things (like subatomic particles) would then derive from particular stances taking up by this 'knowing'." The knowing capacity and what is known (the structure and functioning of the body), as well as the insentient physical basis for knowing, could all be the result of a higher-order knowing.

Our usual understanding makes knowing an exception in an unknowing world. Only the self is allowed to know. But this picture can be thawed. "It is possible to allow a greater measure of liveliness, clarity, and intimacy with our surroundings to replace the isolated appearances of 'body', 'mind', and 'world'." To do this, the next exercises call for reorientation in regard to familiar objects of knowledge. This involves visualization, but not "a mere 'exercise in imagination'."

Exercise 4: Just Interactions and Shining Outlines: Open the interacting structures and regions of the Giant Body so that they are transparent. "Allow the particularity and precision of the surfaces to remain, but as shining outlines, translucent boundaries and interfaces." This counteracts the usual emphasis on stable and opaque things.

"Often 'things' are summary terms or abbreviations for the possible ways of viewing (and things to be viewed), which show what we are actually interested in encountering or considering at a particular time." Exercise 4 lets us develop a mode of seeing not limited to a particular position or point of view at all. We start out with imagining, but a different and incontrovertible knowledge arises: knowledge of how the conventional world of appearance unfolds.

> *Exercise 5: Released to Space*: First make the outlines of Exercise 4 alive and ecstatic; now open up the lines. "The body is gone, released to an intensely alive—yet completely open—space." By exploring 'things', this exercise arrives in the end at a space without qualities and presuppositions. "So perhaps 'things' could actually be said to be, fundamentally, nothing but this 'space' or openness."

> *Exercise 6: Opacity versus Translucency*: See the opaque structures of the body as also being space and wholly open.

The next step is relating to the Giant Body as an embodied person (the psychological and lived dimensions), and also taking into account our own role as observer. Personhood conventionally depends on preexisting time and space; a different view would not accept a prior world, without denying it either. Out of an initial uncommitedness, tendencies develop into feelings and images, which make possible actions and interpretations, which give rise to a "consolidating thrust" that yields the whole pattern, "person in world." There is a freezing of what 'was' (in an atemporal sense) fluid. The result is the usual existence claims. On this view, the person is not a continuous being, but "a tendency toward consolidation or birth in (or as) what for that world are successive moments. The 'person' is then a summary notion describing the overall attempt to set up a frozen pattern."

Compare three non-standard views: a) the world and its uninterrupted linear sequences represent an arbitary, stubborn pattern that derives its energies and possibilities from a more open space, which is then shut out; b) the pattern is only a patterning, which never becomes consolidated, but represent an ongoing tendency; c) no freezing has ever set in.

> *Exercise 7: Body-Mind-Thought Interplay*: The aim is to investigate "you in that situation."
>
> *Exercise 8: The Translucent Person*: Open the partitions between body, mind, and thought, proceeding through the stages of transparentizing and then opening up, as in Exercise 4, arriving at "a kind of openness or 'space'."
>
> *Exercise 9: Participation as Observer; Participation as Embodied Person*: Repeat Exercises 7 and 8, but with awareness of yourself as observer. Include such elements as 'opening up,' 'translucence', and 'space'; then open "the subtle locatedness of this final experience as being the outcome of a certain process," together with its status as an experience. Do not link this openness to more standard world views. When you do "come out," watch this coming-out process.
>
> *Exercise 10: Participation and Space*: Is the observing self "given together" with the Giant Body in its opaque forms? Observe the tendency to 'come out' as a solid, continuous, independent self. Deactivating this self-centered knowing requires sensitivity to a new space-time interplay: a beginning approximation of Great Space.

It seems the space we discover when we look at the Giant Body must ultimately be imaginary, "and that there are no real encounters or discoveries possible." It is true that initially the space we find is only a visualization. But there is a knowing we can trace, and this leads us to the 'space' that is our main concern, which is not at all imaginary. "A basic principle of our view is that anything involved in our being here in the world is directly knowable, not out of reach. And, it can always be traced back to 'space'."

The unity of fixed things and the openness of Great Space is a dimension you know by direct participation. "We can discover ourselves to be that openness." It is not that things are a limited version of Great Space. Nothing has been lost or exhausted. When things are opened up, they go 'into' space. But space itself is not usually considered to be located anywhere; hence things also go nowhere. This is a lower level point, but it mirrors the experiential progression so far.

Chapter 3 49–68

The Mind and the Origin of Appearance

TSK rejects causality and 'source from'. In particular, we need to question the self or mind as knower who has experiences. This assumption is precisely what obscures space. The self assumes a position [cf. LOK] in order to preserve its status; it must then freeze that position. From that it discovers external space (a notion that is true only provisionally). But space is more active than this. "'Space' will henceforth refer to a dimension of reality whose openness is both a requisite and a concomitant feature of all experienced happening." This is related to the 'no-source' view. It gives space as the experienced interactions between self and objects. An intermediate version of this view is discovery of the space that accommodates mind and thoughts.

> *Exercise 11: The Source of Thoughts*: In looking for a source, the point is not to arrive at an answer, but to investigate. What is being challenged is not physical partitions but lived presuppositions, of which 'mind as source' is the most basic. In contrast to certain moves in philosophy, the Ex. 11 experiential inquiry can be seen as 'establishing' that no source exists without that becoming a theory of how things are.

With more awareness, we see that events and thoughts are not a continuous stream. This means we do not have to give energy to making everything fit. "More subtle knowing capacities may emerge, which are not indexed into any 'self'."

This special knowing can see new factors, even intimately embracing the self. Development of this knowing capacity amounts to emerging familiarity with a 'higher space.' Thoughts are meaning bearers, and this gives them the dynamism that manifests as 'from' and 'to': as subject pole and object pole.

> *Exercise 12: Space Between Thoughts*: We may not be ready to confront directly 'here' as being Great Space. But we can contact a higher space "round about" each thought. Seeing the space between is not itself a thought, because the knowing is not appropriated by a self, nor is it a specific event, nor does it rely on the persuasiveness of meaning or amount to a particular and special kind of mental event. The self must relate to its object, thinking its thoughts, but Exercises 11 and 12 "locate an alternative 'seeing' alongside the occurrence of mental events. This alternative is a break in the flow of self-saying thoughts, a profound silence, an openness."

> *Exercise 13: Thoughts as Space*: Exercise 12 is provisional; looking at thoughts as such, the same freedom from subject-object fragmentation is available. It is Great Knowledge that understands this.

What is 'meant' by the suggestion that there is no mind? "Meaning" is not the right issue, since this is no ordinary proposition regarding existence or non-existence. Instead, the focus is on experiencing "no mind as generative source," on the discovery that "mind, like space, has no foundation." "The mind is a sign of a particular focal setting being taken on Great Space," and this helps explain why we do not find it. Mind gives rise to all appearance: The exercises show this, and also help release it. "This is accomplished by working with what could be called outputs of the 'mind-setting' *in ways that are uncharacteristic for that setting.*" Mind stands in a two-term relation to its thoughts, but Great Space does not stand in a two-term relationship to its manifestations.

Exercise 14: A New Focus on Space: Repeat the Giant Body series in light of later exercises, with a special focus on the subject pole. "Experience the insight that mind is not a rigid and continuous thing, but only an arbitrary setting."

Chapter 4 71–88

The World and Other Bases

By extending Ex. 11–14 to the body-mind-thought interplay of Ex. 7, we can complete our review of what it means to be a person. Then we can challenge the highly structured world, which will prepare us to move toward transcendence of the present "closed circuit." Thus, as an extension of "no source from," consider this: no by-standers; no outside-standers. Key to this challenge is giving up the search for a first cause, because this approach commits us to setting up one standard 'world at large'.

This is not to deny divine dimensions, but to challenge the way they are adapted to conform to a particular lower-space orientation. Great Space as ground counteracts cause and effect perspectives, "which constrict the vital and infinite dimensions of immediate appearance."

Great Space is not elsewhere, but it is also not 'here'. 'Here' is a distracting bystander, tied up with various outside-standers. To say that Great Space is 'here' is simply to "tone down" the nearness of infinity to a level tolerable for a 'self'. When this happens, "[t]he infinity of Great Space . . . unfolds as a particular world, an indefinitely extended field."

Causes, whether scientific or mythical, only reinforce the sense that we are located in a spatially and temporally extended environment, so that we cannot appreciate the immediacy of Great Space. While this restriction is characteristic of religious views, religion plays a useful role in challenging the most restrictive forms of self-orientation. At higher levels, such approaches are ways of dissolving limiting focal settings on Great Space. The techniques of religion, which open focal settings, thus contribute to the quest for knowledge in the physical sciences. In this way, Great Space inquiry can reconcile science and religion.

The approach till now has used many of the features of the familiar settings, but now we can look at how to "let go of the output of the focal setting, without this process establishing that the output ever existed in the first place." 'Here' turns out not to be so fundamental after all. Great Space makes it clear that the profound and infinite are not located at special points in lower time, such as beginnings or ends. To see this requires deactivating 'goal and progress' orientations.

Since everything that constitutes existence is the product of a focal setting, everything (including the self) is alike in being 'given together'. Some examples of views that oppose this insight at subtle levels: idealism or mentalism; relativity; various forms of idealism; sense data as real; and the meditative injunction, "Be here now!"

Review the freezing thawing processes of Ch. 2 and the transparentizing introduced in the exercises. The earlier sense that these exercises are done by you, here, needs to be deactivated. "By challenging and opening up the output of the setting, working with one aspect or another of this output, we can begin to change the focal setting itself."

This has to do with arriving at the 'given together' insight, which is related to Great Knowledge (while the togetherness itself is related to Great Time). First we need to look for subtle aspects of self and realm, then challenge them experientially, and then penetrate their partitions, as with the Giant Body. That is the point of all this challenging, for after all, we are not doing science, philosophy, or logic. Finally, we let the partitions be, embracing and releasing them.

By transcending common appearance, or by appreciating Great Space as that appearance, we arrive at Great Space. First we are set free from the cave; then we learn there is no cave.

Chapter 5 91–114
Being in the World, Being Space and Time

Attainment orientations, based on notions of 'person' and 'world', are too fixed to arrive at the open, liberating dimension of space. Even the notion of 'experience' is too limiting. Challenging this is not a preliminary; it can be the path to Great Space. Every partition is a doorway to higher spaces, so it is good to identify and challenge all such partitions, including the presuppositions involved in challenging. Attaining a wider perspective lets us disengage our attention from "subtle but influential forces structuring experience."

We need to integrate the depth and width dimensions of previous chapters:

> *Exercise 15: A Mountain Retreat*: Breathe in blue sky. Let the space pervade all aspects of experience; let 'you' and sky commingle, in conjunction with breathing. Surrender all positioning, without negating anything. Let go of directedness and locatedness and the fragile 'here' associated with getting. Also release 'experience'. Eventually, look for this same experience in the hectic city. The energy available here is not based on goal attaining or grasping. Ordinarily we say the world grounds meanings. But here we see that the world *is* meaning, "a being 'given together' as a 'meaning'." (97)

Now comes a transition to time as the second facet of 'existence': the dynamism that Great Space potentiates. Great Time "conveys" Great Space in both senses of the word. When a restrictive type of 'knowing' judges Great Time's evocation of Great Space to be things and existents, "it becomes necessary to assert lower time and space to be in force." At this level, time seems spread out or distributed. It times out things, differences, qualities, etc. The meanings of such a knowing support this 'drawing out', which makes 'here' dependent on 'there', on 'up ahead', and so on.

"The 'going' of time is due to its dealing in terms of meanings and existents." Existents are inherently deadening, while meanings require a static position defined by static surroundings and containers. Boundaries, partitions, and limits must all be set out. But all defining surfaces are still incandescent and malleable. (This holds for the self, but that is too threatening. The self does not want to get outside lower space.) Meanings are the echoes that sound out the walls of the confining chamber, but in fact they define those walls as part of the given-together. Meanings allow the walls of 'things' to communicate with the walls of the chamber, but only at the cost of friction.

Non-existence in the Great Space sense is not opposed to existence, which is a term that arises only in terms of a linear series related to causes. Non-existence in this context refers to gaps. But there may not be gaps, series, and causation. "Appearance does not come *from* something, nor is it offered *to* anyone." "Since Great Space is open and accommodating, the lack of inhibiting influences is itself sufficient for appearance to *be*." Time is not intrinsically transitory. Transitions are just a way to shake off superficial constructs and clinging. The series is a fiction: There is no from or to, no creation, no existence. We live in a magical world, with no one performing the magic. We make this magic law-like and rational; or rather, we *are* the tame, law-like rational 'seeming'.

> *Exercise 16: Space-Time-Knowledge on the Conventional Level*: Let all positions and forms be given as space, all units, meanings, and actions be given as time. The two carry a knowing dimension, without a knower. Relax locatedness, locating, here, now, happening, etc. Release all bystanders and outside-standers. You can relax the weariness of insisting on existing.

Three stages of space have been discussed. *Level One* is precisely our ordinary realm, seen in the light of further possibilities of human fulfillment. (111) All observations are seen as the function or output of a local limitation on Great Space. This lets us see all appearance as pervaded with the Great Space dimension, which is not spatially distributed or extensive. Each finite point has infinite space, and no one thing excludes any other thing. Nothing has been set up, and there is no one standard world order. *Level Two* comes when we make experiential progress in opening the focal setting. Now we are living in a second-level relationship to Great Space. But there is still a subtle, self-centered or 'standard world order' interpretation; we still appropriate the freedom of Great Space. *Level Three* involves the breakdown of levels. The strategy of opening focal settings collapses. There are no lower spaces, and we transcend the self-centered orientation. All is a great play.

Chapter 6 117–32

The Presence of Time—Liberating Potency and Compelling Patterns

Theories are the expressions of 'hidden' forces or dynamic principles. If we understand these forces, we may be able to transcend our control and improvement orientation. The traditional way to understand events was in terms of powers possessed by objects. Now we look at law-like sequences. The next step is for these regularities to be seen as "simply an expression of the structure of time." Science can deal with sequences of events, but has no way to investigate the sequential nature of time itself, to ask why anything should happen at all.

Everything in our realm depends on the 'flow' of time: that something will happen next. Seeing reality in this time-flow sense, turning from the focus on objects, raises the possibility that objects, their states, and their interrelationships may somehow be a function of time itself. Normally we stay focused on objects, and consider time only an abstraction or index. As a result, we seem unable to develop accurate theories or to realize the full value of our lives. There is some subtle dimension of time we are missing, due to our self- or agent-oriented picture. Lived time in the ordinary sense is not just a hidden force that pushes us around, but our sense of helplessness is a sign that something vital to our being is being left out of account.

We can approach time in terms of trends and regularities, gaining more power in terms of both psychological and physical dynamics. This is a way to work within a particular 'lower time'. The first step is to understand the lower time, that operates in our realm: the 'local', dynamic expression of our lower space. Unfolding in this way, time gives isolates that interact and can be indexed in terms of past, present, future moments. Time distributes itself across ordinary time, and the totality of Great Time is made manageable as the next moment, transitoriness, and so on. But even the shutting out of the whole derives from Great Time. This does not mean that time is subjective, because the subject is also a result of a particular 'knowing' of time. But the resulting ordinary time does conform to the ego's concerns and restrictions: time flows in an orderly way. Even on this level, the movements or series of time does not just *confirm* the identity of a thing; they constitute that 'thing'. "We are literally timing ourselves away," ceding control of our lives to 'time'.

When we break through to Great Time, we go beyond energy and also beyond 'subjective' and 'objective'. Time and Space become the central facts, so that "each presentation, each unit of ordinary time, bears the possibility of contacting Great Time." Now we can control the direction of lower time. "Time can support as much experience and living as we like." We can end our reliance on technology, psychology, medicine, religion, etc.: "We can accomplish more than any human beings have throughout history."(132)

Chapter 7 135–62
The Path from Ordinary to Great Time—An Overview

Ordinary and non-standard experiences of the flow of time are just transitional, and conform in some sense to conventional presuppositions. But 'flow' is a good starting point, even if we then move on to the view that there no moments and no succession. a view that can be supported not just theoretically, but with a new way of knowing.

Stage One in the appreciation of Great Time recognizes time as an abstract index and a compelling force. This mode of time is the ego's attempt to tame Great Time, and it expresses the infinity of time as the ongoing, endless flow of moments. We keep time in the background, and it manifests only in the interaction of objects. Attempts to break out of lower time (e.g., through meditation) hold to the familiar character and logic of things: We may just produce fantasies, or else freeze time briefly. When we settle on 'things,' we lose a measure of accuracy, and our ability to clarify what is actually happening will suffer. But "thinking can imitate 'time's' penetrating evocation of Great Space." The *penetrating* investigations in the Giant Body exercises suggest some of this: a "deep thinking and an intellectual transcendence" of isolated 'things'. (141)

Stage Two involves experiential acquaintance with time. Even a slight intimacy with time shows it to be enabling rather than oppressive. This calls for a different 'Knowledge'. This experience is what brings the TSK vision alive. Initially second-stage experience may show time to be like a flow or flashing, due to our making it conform to the usual serial progression. But we can learn to allow brief 'knowings' that take into account all that appears, including ourselves, as being given by time. There may be a sense of improvement that summarizes 'time', just as objects and identities are summaries of time sequences. A certain degree of higher knowing is left unobscured, and we appreciate 'things' as symbols. We can focus on the present without tying it to its before-after causal nexus. This then unfolds into a sense of time presenting: a 'timing' that gives all appearances together, as well as appreciation for those appearances.

Second-stage time lets moment give way to moment. At the same time, time lets specific items have their identities, while letting those identities evoke the 'no-things' of space. Insight into 'timing', with 'things' as timed-out meanings, give 'weightiness' to no-self critiques.

Finally, time at this level is the bridge to other realms. We discover space-like gaps and discontinuities, and because our transitional concern is with 'breaking out', we naturally focus there. But we can also proceed differently, skillfully "feed[ing] 'time' back into 'time'." (149) This opens knowing into space, in a transforming alchemy. Instead of moments succeeding moments, we see that each moment in the series can be 'timed out' in such a way that it leads into transcendence.

Eventually, we may see all timing as happening in one point. A 'non-going' results in a collapse of here/there, allowing and celebrating Great Space. There is a growing emphasis on intimacy. We discover (153) that every point can be interacted with (even if this is not so on the conventional level), due to the presence of pervasive space.

Seeing things as space and time, "we can move and open through these dimensions to expose infinite fields of possibilities within what seemed a finite encounter." (156) This 'can' becomes a 'should', implying responsibility to act on what is appreciated. But if it leads to "the feeling of intimacy or mutual interpenetration on a cosmic scale," (158) it can become a trap. What continues to cause confusion here is the conviction that *there is something*, as well as a sense of 'occuring' or 'happening'.

Stage Three time is Great Time, which is neither law-like nor random; not a happening 'taking place'. It is not a mysterious force. It offers a complete and uncontrived intimacy: the surfacing of our real Being. Although we are free to take different views, Great Time and Great Space join in marriage, yielding the appreciations of Great Knowledge as unending consummation. There is the awareness or image of never having strayed, despite an eternity of straying. There is a fluid movement, a perfectly, liquid, lubricious dimension, with no going and no separate places.

Chapter 8 165–90
Intermediate Experiences of Time

To move toward a level-two experience of time requires familiarity with Great Space. The relationship is not theoretical: In space exercises, it is possible to trace the emergence of 'timing'. It is also important to develop a wider perspective on the self, which cannot understand Great Space or Time "because it is precisely the *embodiment* of a lapse of such understanding." (166) Still, as a patterning of time, the self bears knowingness, which can be contacted.

The self declares itself in charge, but this is a move made by Time (or by 'time', its stage hand.). A local limitation occurs, and time becomes finite. Gradually a consolidation 'takes place', based on a polarization that gradually turns the infinite into "a facing off between two isolates in a finite slice of ordinary time . . . The power of Time fragments, becoming the volition of the self, on the one hand, set in contrast to the energy of nature and physical processes, on the other." (169) The self and objects are timed out by 'time', but this 'timing' is lost from view. There is growing diffusion and fragmentation, resulting in highly charged dichotomies and a sense of distance and separation.

> *Exercise 17: The Object and Its Glow*: Visualize the self as the object's glow, and reflect on the ordinary experience of distance. When we stop moving through distances, a new 'space' becomes available.

The self inhabits an impoverished present, as seen by the fact it is always going somewhere:

> *Exercise 18: Past and Future Projections*: Notice how much we dwell in past and future.

> *Exercise 19: Past, Present, and Future of Each Moment*: Find the past and future tinge to each moment. This can be a knowing that the self does not do.

> *Exercise 20: Reversing Temporal Structure*: Look *from* future or past to present, or from both simultaneously.

> *Exercise 21: A More Subtle Structural Reversal*: The value of such exercises is to counter subtle or overt *temporal directedness*.

> *Exercise 22: Diving into Time:* Go from future to future of moment to moment; past to past of moment to moment, etc. Eventually it is possible to find an infinity of nested past-present-futures. Observe especially the doer in each of these moments. Working with this always-emerging self greatly reduces the usual inertia of the self.

Through such exercises, we learn to contact time more directly. This is like discovering moments between moments in the temporal 'points' of the zigzag diagram (179), though in a third-level perspective there *are no points*. And since we are 'in' each of those inter-moments, we can discover more points, and so on. Since time lines are just constructs, we may in this inquiry come upon the initial points and sequences.

What makes it possible to move along the 'time line' is actually an ignoring. But now we can gradually return to Time, shifting from the zigzag to a smooth line, which becomes in turn a point, a centered abiding, and finally a centerless center, a 'no happening' that can expand everywhere.

> *Exercise 23: Going without Going*: A very slow, stylized walking can develop a 'knowing' more broadly attentive than the self. This may be threatening (for the self), but through it experience can expand.

> *Exercise 24: A Marriage of Sound and Breath*: Focus on the breath and the communicative aspect of time, and also on the qualitative presence of sounds. Marry the two attendings. You can cultivate a knowingness not deceived by meanings.

Exercise 25: Intimacy: Work with the recognition, "That too is 'I'." We can progress from artificial to uncontrived intimacy, and then to intimacy that involves neither subject nor object.

Chapter 9 193–208

A Self-Challenging Vision Opening to Great Time

Our ordinary encounters, reactions, etc. actually require us to ignore time; to treat it as a stable background within which meanings and identities form through a process of referring. The 'timing out' model challenges these stable identities and meanings. This is the read-out principle: no situation relates to any other. No point even establishes that there will be a next point, or establishes itself *as* a point. Every message as to "what is so" is completely self contained.

Due to our forward and outward directedness and our need for continuity, successive read-outs seem closely related. But this seeming reality can be doubted. It is not that 'we' can walk through walls, but there could be such an "uncharacteristic" sequence of time. "Sameness is a matter of conviction *within* a read-out . . . Seen in this way, all existence and experience is like an apparition, a surface with no substantial core." (198–99) But this too is at best a transitional insight. No read-out establishes anything beyond itself, nor does it establish itself. There is nothing that cannot be challenged. Seeing this is the transition to Great Knowledge.

Incomplete insight into the read-out principle can lead to the claim, "all is illusion" or "all is mind." But this is to stop too soon. Since everything is open to challenge, different claims and understandings (e.g., religious claims) need not contradict each other. Nor is relativity established.

Exercise 26: Transcendence of Pointings: Investigate all that appears in the mental and physical realms as mutual pointings. All presentations amount to a saying that says nothing, undermining itself.

Chapter 10 211–16

The Possibility of Knowledge as Commanding Insight

Space provides limitless opportunities, and Time can transform everything. Both make possible Knowledge, which is both the goal and the basis of the path. "Great Knowledge is the immediate and knowing dimension of all reality and experience . . . Knowledge is not something which knows something; it is simply the presence of reality as 'knowingness'." (211–12)

From a lower-level perspective, there seems to be a separation between Knowledge and knowledge, but that is an ultimately self-limiting approach. Instead, it is just a matter of living out the unfolding vision. "When Space and Time are directly appreciated, that is automatically Knowledge." (216)

Chapter 11 219–30

The Obscuring Nature of Ordinary Knowledge

We see the world as insentient, and therefore can only act in it in clumsy ways. But we can thaw that world. This is not a theoretical statement; in fact, it "can be brought to life precisely by examining the basic presuppositions of all such theoretical knowledge." (221) Such a critical examination is indeed vital, for it shows them all to be variations on a theme.

A critique of lower-knowledge points out that we know only our own experience, but that is not a telling critique in lower-level terms, since it is cheerfully conceded. "It is philosophically unsound to push the significance of these limits very far." (225) However, it is valid and useful to cultivate the *feel* of such a challenge as an interim measure. It helps us realize that all that we know can be resolved into a few standard 'mindings'—just images and concepts.

Human affairs in history have been run largely on the basis of beliefs, especially the belief that something is lacking. Other beliefs are put forward about supposed solutions. But the result is a subtle coercion, a closing down of intelligence that also robs us of time and space. We have been the victims of such manipulation. Now we have to act differently, penetrating all partitions. We have been perfect believers, but now we must awaken knowledge, intelligence, and appreciation.

Chapter 12 233–48

Lower Knowledge—a Hindrance and an Opportunity

The 'minding and belief structure' trend is particularly objectionable in not readily allowing critiques of itself to be made. It solicits our energy and attention in a way that keeps us trapped, convinced there is no alternative. To counter this, we need to challenge this trend in ways that preserve *none* of the basic terms and presuppositions of ordinary knowledge. Looking at exactly the same content we already know, we can say that "it is all just 'lower knowledge'." (235) This formulation successfully deals with objections to conventional critiques. Lower knowledge is just a more comprehensive way of comprehending what ordinary knowledge deals with.

On this view, physical objects are themselves just knowledge, without this claim amounting to a form of subjectivism. Instead of subject and object, a particular and constrictive 'space', 'time', and 'knowledge' operate: The world is a meaning or belief, and we can truly challenge everything. Even what seems creative is seen as repetitious, like a recording. Everything must conform to an implicit logic that stays on the surface, engaging only lower time and lower space. The knowledge allowed is unreliable and unfulfilling. But we have been afraid to question, sensing that "we might end up with 'knowings' which go far beyond the scope of the self's territory." (239)

Great Knowledge lets us go beyond theoretical challenges to knowledge. It supports a "felt confrontation" which shows that the account lower knowledge gives of itself cannot be right even in lower-knowledge terms. Experiencing lower knowledge *as* lower knowledge lets us begin to access its 'knowledge' component more directly. We can drop the picture of 'something' knowing 'something', in which isolation is primary. Instead we can attend to 'knowings' as such, "without 'bottoming out' in terms of 'things'." (241) In this view there is no 'not-knowing'. When we freeze knowingness, we cut ourselves off from this alternative. "As long as we believe in a discrete and solid world—and lack the awareness that there is only Space and Time—" (243) we weaken the fabric of our world and choose decay.

Our usual diversions undermine the possibility of being mindful at this deeper level. Just sitting quietly is an existentially charged action. "The lower space we live in . . . changes, *becoming more or less open, with each single thought* . . . Second by second, mindfulness (knowingness) must be maintained." (244) This vision is both open-ended *and* rational.

The range of concerns in any particular field of knowledge—e.g., science—can be considered a cone. Science cannot know what is beneath, or around, or within. Religion at least acknowledges that 'transcendent' region, but maintains that it is closed to thought or inquiry. But Great Knowledge disagrees. "It turns out that each ordinary point [*is*] that transcendent region." (248)

Chapter 13 251–60
The Emergence of Great Knowledge

Everything can be seen as Great Knowledge. The point is not to force special experiences, but rather simply to relate to what Time and Space continually display, until we arrive at the Body of Knowledge.

The ego wants to control, but there is no need to control. To see this, we can seek out those elusive moments of clarity and significance—a balanced encompassing.

> *Exercise 27: A Stream of Memories*: Remember remembering, and repeat: an unending series.

Exercise 28: A Cycle of Seeing: Choose ten things and cycle through them mentally.

Exercise 29: Awareness as a Reflective Surface: Allow all objects to be known; then de-emphasize 'presenting to' and 'yours'.

Exercise 30: A Subject-Object Reversal: Do this first with objects, then with thoughts and experience. These exercises show knowing as coming from a different level, though they make concessions to knowing as two-term.

Exercise 31: Not-Knowing as Knowing: Within insentience is knowingness; there are no 'dead spots'.

Chapter 14 263–78
Great Knowledge—Its Value and Pervasive Quality

Knowingness can maximize love and intimacy. We would like to enjoy the present, but "we suddenly find ourselves confronting *subsequent* situations." (263) But even when knowingness lets us cultivate positive aspects, the 'motion' of 'time' leads to the 'coming out' tendency of the self and pointings. As a drawing apart tendency and consolidation, the self is naturally unbalanced, naturally grasping. For instance, in love there are three layers, the second being impatient and demanding and the third more neutral; in anger, the second layer is incisive and penetrating. Through such patternings, our personalities form. Instead, we can tap the knowingness within the labeled experiences, discovering that everything is 'within'. We can be 'in' the energies we are. For instance, in love there is the hunger of the consolidating self, which draws apart from experience and hence finds itself craving, so that we lose the nectar of love.

Nothing *conveys* the clarity of 'knowingness'; there is just the clarity itself:

Exercise 32: A Unifying Clarity: Mandalas of lights, whose light 'carries' knowingness. Repeat at a more subtle level, until there is only knowing.

Exercise 33: Self-Transcending Appearance: Let every appearance reveal freshness, clarity, and light.

Exercise 34: The Embodiment of Knowledge: Light and knowingess within embodiment; the body in its knowledge aspect.

Exercise 35: An Evocation of Knowledge.

Chapter 15 281–90
Knowledge and Totality

Open-ended clarity, rather than mind or things, can be our orienting guide. Knowingness does not radiate from a center, and things are not separated in time or space. The darkness imposed by the self can be illuminated and then recognized as inseparable from light.

There are three levels in relating to this knowingness. At the third level, there are only Space and Time (but this is not a comment about reality). In first-level terms, we could say that everyone participates equally in the Body of Knowledge, leading to uncontrived intimacy. Unification comes about through incisive clarity (knowingness) and the linking capacity of time. We arrive at zero-dimensional being.

"Statements about Great Knowledge do not take up a significance alongside ordinary models and presuppositions, but apply to the alternative and comprehensive Space-Time-Knowledge vision." (288) We begin to relate to this vision through a critique of our own mind, and then extend that critique to open to knowingness, so that mind loses its defining edges. We realize that ordinary mind did not exist in the first place, just like ordinary objects and subtle patterning tendencies. Whatever appears participates in an "an ongoing self-transcendence." (289)

Chapter 16 293–306

Knowledge, Being, and Human Being

Living the Space-Time-Knowledge vision lets us find the way to Being, where these three facets unite and manifest in the comprehensive intimacy of the Body of Knowledge. In appreciating Being, we arrive at an understanding of why Space, Time, and Knowledge are as they are, and why just these three are the facets of Being. "Within—or *as* presence, Space, Time, and Knowledge shine in their respective ways " (294)

Talk about Being makes sense only at the second and third levels; attempts to use relevant terms at the first level, as in religions and metaphysical systems, remain ungrounded.

The self does not arrive at Being, which is the paradoxical presence of appearance: presence as wonderment and sheer art. All actions are profound ritual gestures and also cheerful, lighthearted play. We begin by celebrating and nurturing. Then we discover that we all hold the same value and kinship, as shown by our all being within the same cycle of time. We arrive at the insight that one point is all points. Self-seeking is impossible. We are ready and able to be of benefit. Fulfillment arrives spontaneously.

Love of Knowledge

Chapter 1 3–9

The Promise of Knowledge K1

Knowledge is fundamental to who we are, but routines blind us to knowledge. Conventional knowledge can be seen in the light of the possibility for higher knowledge, with which we already have experience. "Step by step, knowledge could teach us knowledge. We could pursue questions in a new way; not just to accumulate information and ideas, but to learn who we really are and what we can do with our lives." (5)

This knowledge of who we are is the real knowledge we seek. To pursue it, we have to develop our ability to question; we have to find the knowledge within models, visions, and ideals. It all depends on inquiry.

Chapter 2 11–15

Knowledge Alive in History K1

We can look at the past and at how knowledge manifests. We can make knowledge the topic of inquiry. Appreciating knowledge itself a form of knowledge. "The source of our knowing merges with the subject under investigation." (14)

Chapter 3 17–23

Faces of Knowledge K1

In the history of knowledge, technology has now come to the fore. The result is to make knowledge constantly change, encouraging us to ignore the past; a further result is to make knowledge homogenous. But whether this is the best knowledge is open to doubt. "New possessions, advances in science, or even better health and longer lives may mean little if we lack the knowledge to choose our actions wisely, with a full understanding of their consequences." (21)

Chapter 4 25–31

No Direction K1

Technological knowledge gives an understanding of what knowledge itself is. But it does not give a way of questioning that understanding. "Technological knowledge leaves its own way of knowing outside the scope of inquiry." (28)

Chapter 5 33–41

Binding Through Identity K1

The value of the results technological knowledge yields is not a valid question for knowledge. The technological model divides the objective world of results from the subjective world of personal conviction and concern, where knowledge simply does not apply. "Since issues of value and meaning fit into the subjective realm, they recede from view as possible subjects of knowledge or topics of public discourse. In such circumstances, what is meaningless comes to the fore by default." (33)

This split is reflected in the model of the self. The self 'uses' knowledge. But the identity of the self—indeed, the fact of identity, "to which all knowledge is subordinated" (34)—is inaccessible to knowledge. The subjective/objective split gives rise to desire and then action, guided by technological knowledge. But any deeper knowledge is cut off. "Knowledge becomes a second-class citizen in the self's subjective world, wholly subordinate to the structures of desire . . . Such knowledge does not directly affect the self in its being." (36–37) As "subjective," knowledge is denied real significance.

Chapter 6 43–49

The Known K1

Technological knowledge is shaped by labels and identities, by what is already known and established. The result as we experience it is that life loses its vitality. But it does remain possible to investigate the mechanisms of limitation.

Chapter 7 51–55

Replacing Knowledge with Belief K1

We tend to replace the search for real knowledge with the acceptance of a belief, and these beliefs work to cut off access to direct knowledge of experience. We do this based on the belief that we ourselves lack access to the knowledge we need. Beliefs arguably carry knowledge, but we have no access to that knowledge, because we are too caught up in our limits. We accept a substitute for knowledge that will give us no access to knowledge, *unless* we can find a way to discover for ourselves the knowledge the belief embodies.

Chapter 8 57–65

Questioning Limits K1

Accepting beliefs leaves us feeling helpless and limited. But we can take these feelings as a warning sign that the knowledge we rely on is incomplete, and look for ways to change this. "The possibility emerges that knowledge can go *behind and beyond itself.*" (59) We can start with the facts of our life and mortality and our endlessly frustrated search for dominion. We can acknowledge how firmly established the patterns that restrict us are. Whatever our limits, the possibility remains of seeing the patterns in operation.

Chapter 9 67–73

The Range of the Knowable K1

The knowledge we rely on does change, and whenever we trace its origins, we come up against a blank wall. Even our most fundamental convictions—including the conviction that we are bound by our patterns and our past—may not be well established. There are many ways to awaken a sense that knowledge could be more than we ordinarily conceive. Conventional knowledge wants to affirm its characteristic not-knowing, but we can see that more is possible. "Even within not-knowing a fundamental clarity is at work." (72) Seeing this, we can begin the work of inquiry. "The prospect for transforming human destiny may already be implicit in the first real question we ask." (72)

Chapter 10 75–81

Adventure of Knowledge K1

Open questioning has its own rhythm we can tune into as soon as we are willing to observe. For instance, we could question the self. Instead of being bound by positions, we can "simply be aware whenever positions are in operation." (76) We can take responsibility for pursuing a deeper knowledge. We can turn from the model that makes knowledge into *private insight*, for the knowledge we seek is not private. We can find joy in the possibilities.

Chapter 11 83–87

Disclosing New Potential K1

There is a way to look without relying on established positions. We have to challenge our model of knowledge and also the accepted rhythms of knowing. [Knowing is an activity, and recognizing this leads us to investigate time.] "Space and time themselves are transformed through the infusion of knowledge into all appearance." (86)

Chapter 12 89–95

Transforming Time and Space K1

The time and space that lower knowledge offers have well-known limits. But deeper knowledge transforms these limits. We must let go of preconceptions of what will benefit the self. We must acknowledge that we do not know that we do not know. Then "we find knowledge less in the content of what we know and more in the activity of knowing itself." (90) We learn to evaluate the quality of our knowledge. The changes that result may not require linear time; may not arrive in the future or require breaking through barriers; may not be owned by us. Knowingness opens.

Chapter 13 99–107

Knowing Through Polarity K1

Space is regarded as empty; time is dismissed as the grid for measurement; knowledge is held to be based entirely on sense experience. This is temporal knowledge, linked to a particular time and place. Temporal knowledge is polar knowledge, grounded in subject and object. It starts from not-knowing, and must transport knowledge across a gap: from 'there' to 'here' and 'then' to 'now'. The movement toward knowing rests in turn on the distinction between self and world. The self generates descriptive knowledge, labeling and identifying a known world. Descriptive knowledge draws on the past: it sets in place a perspective with a fixed focal setting. "To define is to be confined within the definition's narrow range." (105) Such knowing is always drawn away from experience, so that not-knowing always remains primary.

Chapter 14 109–19

Problematics of Polarity K1/T1

The structure of polar knowledge is inherently untenable, and the consequences it generates are unacceptable. For one, the knowledge we really need is unavailable: polar knowledge leaves us restricted. "Investigating polar knowledge means asking whether we are free." (112) For another, the connections necessary to sustain it cannot be made: linear time proves incapable of supporting knowledge at all.

Chapter 15 121–27

Compelling Flow of Time T1/K1

There is no way to account for the unfolding of linear time, nor is there any way to account for the first moment and other anomalies. This means that there is no true basis for descriptive knowledge. And if a basis *were* posited, it would depend on the unverifiable accuracy of memory. "It is only *by stipulation* that we can say the echo is an accurate one." (126)

Chapter 16 129–35

Modeling Self-Constructs K1

Descriptive knowledge cannot give us experience. Then is it knowledge at all? When pressed, it cannot prove its claims. We have a model of knowledge that says that knowledge comes through models. But models cannot validate other models, nor can they appeal to experience. Nor are models justified on pragmatic grounds, for they are not inherently successful: they leave out change, cut off certain kinds of knowledge, close off experience, and shut down inquiry into fundamentals.

Chapter 17 137–45

Interpretation at the Core K1

Governing models say that our models model an underlying reality, but in the end the only justification for this claim is other models. Constructs match up with other constructs, interpretations with other interpretations. The real justification seems to be *the witness*: "the 'feeling' of reality, or the conviction that 'this is so'." (143) Compare the imagined cup and the seen cup. The usual witness is the 'I'. "Because the self 'exists', the world is 'real'." (144)

Chapter 18 147–53

The World of the Self T1/K1

The self moves out of the point-instant of temporal knowledge. "In one unified action it takes a position, posits a situation, and imposes meaning." (147) It insists that it persists across time. This persistence emerges out of the future (unlike descriptive knowledge, constructed from the past). The self desires to know and establish a world into which it can emerge. "The force of the self's desire unfolds as the momentum of linear time." (149) The self insists on being owner, definer, and shaper of experience. Its relation to what appears in objective space and time is deeply problematic, but the self dismisses such difficulties.

Chapter 19 155–61

The Reign of Intentional Knowledge K1/T1

Intentional knowledge complements descriptive knowledge, making the revealed world meaningful to the self. But it remains bound to desire, and thus assures dissatisfaction. "Tightly bound by yesterday, swept along into tomorrow, the self is cut off from its world." (156) The self lives like a traveler in a foreign land. It tries to turn toward feelings and emotions, but the result is to shut knowledge out entirely. [cf. Ch. 5] or it tries to control its circumstance, generating technological knowledge. "Space is filled up with the structures adopted by technological knowing; time is measured out in advance." (160)

Chapter 20 163–67

Guidance and Direction K1

To counter the anarchy of feelings and the emptiness of technological knowing, we turn to collective knowledge: the persistence through time of intentional and descriptive knowledge. But the result is only to choke off creativity and autonomy differently.

Chapter 21 169–77

Presenting the Self T1/K1/S1

"Just as in the space-centered view of 'objective reality' space itself disappears in favor of objects, so in the temporal view knowledge as such disappears in favor of the intentions and concerns of the knowing self." (169)

The self sets itself up as narrator, telling the founding story of the self. In this way, linear time and a world subject to the flow of time are established. "All knowing is directed toward points after the flow of self-centered experience is in operation." (174) This "single-minded" knowing repeatedly confirms what is so. The past is made the source of reality.

Chapter 22 179–83

Two Times T1

Self-identity is bound up with linear time. But how can the self occupy time? There are two views of linear time that conflict: the 'objective' linear time of cause and effect, and the threefold time of experience, which ultimately takes it structure from the presence of 'I' in the present, and in this sense is wholly subjective. There are similarly two views of the present: one that sees it as zero-point dividing line and the other that finds within it the fullness of all experience.

Chapter 23 185–91

The Onward Flow of Time T1/T2

There is no good explanation for time's flow or momentum, which appears to be what constitutes narrative time, and which seems linked somehow to the vitality of the present. Nor is it clear how self as owner of experience—or as perceiver—can link to this flow. Perhaps the witness in its undeniability points to something more fundamental in the dynamic of time's energy.

Chapter 24 193–99

The Self in Question T2/K1

The witness can be understood as testifying, not to the content of what is so, but to the availability of a deeper time: one that presents all meanings but is prior to all meanings. (193) To arrive at this level means challenging the story of the self: seeing the self as "the shadow image of 'my own' knowing." (195)

Chapter 25 201–209

Founding Identity K1/T1/T2

The self in its problematic relation to time opens up the nature of time. But we hesitate to pursue this inquiry. We prefer to accept linear time and the rule of the past, because it gives us the existence of the self, on which we insist. And so we accept restriction; we bargain knowledge away.

Chapter 26 213–21

Merging with Time K1/K2

Conventional knowledge (mediated through the structure of self as knower and owner) depends on a certain understanding of space and time, and hence could be quite different. This is just what the self is unwilling to consider. Instead, it chooses devotion to history and to what is so. The issue is finding ways to go beneath the 'historical self'.

Chapter 27 223–29

World without Positioning K1/K2

The challenge to lower-level knowledge triggers various defenses: fear, preoccupation, confusion, or appropriation. "The self lives in the world like a permanent alien." (226) Such responses express the logos: the governing logic of the founding order. But confusion and fear can be reinterpreted as the onset of knowledge, signs that the logos is open to challenge. The opening comes through a questioning not done by the self. Knowledge can serve as its own witness.

Chapter 28 231–37

No Alternative in Alternatives K2

The idea of an absolute, whether religious or philosophical, offers an alternative to lower knowledge. But understood in lower-level terms, an absolute is untenable; understood as incomprehensible, on the other hand, it only affirms the limits on knowledge. So making this move is ineffective as a way of transcending level one. However, faith and devotion do have their place as pointers toward knowledge, challenging limits, deepening appreciation, and opening vision.

Chapter 29 239–47

Visionary Realm K2/T2

If we replace the narrative with the vision, we come closer to real transformation. The vision of the robe allows new ways of knowing to emerge. There could be a timeless, dynamic time of no structure. We could experience time directly: its rare creative energy. We could arrive at 'random' time. Yet such visions, in setting forth a higher reality, only confirm the limits of lower knowledge. "The turn *toward* the vision subtly affirms the reality of what is turned *from*." (245) We must return to our present time and space and knowing.

Chapter 30 249–55

Touching the Limits K1/K2/K3

Instead of aiming for a different realm of knowledge, we can trace the source of the limits on lower knowledge. Our usual models aim at specific purposes, but such purposes limit the knowledge available. A 'fundamental model' refers back to the self. The model confirms that something is lacking and something must be done, that knowledge must be obtained. Efforts are made, but friction and failure are the outcome.

Human beings become consumers of knowledge, rather than its producers.

Chapter 31 257–61

Constricting Circle K1/K3

How is it possible to know in a new way, when we only have the old ways available? We seem caught in a web, a circle of not-knowing with no beyond; we seem, even, to consist of our limits.

Chapter 32 263–67

Taking a Stand K1

Our positions (lower-level space) and our conditioning (lower-level time) close us in completely. To break out of this, time and space themselves would have to be different, but this would require a different model of the self. At present, we adopt the bystander model, which proves more basic than the polar model of knowledge (a hypothetical construct). Knowledge is based on positions and oppositions.

Chapter 33 269–77

Availability of Knowledge K3/T2

Reasoning cannot undermine the bystander, since the logic of rationality and the bystander are born together. Instead, we can turn to a more direct inquiry. When "we can inquire into observation as it operates," (270) labels and ideas lose their power to confine the range of inquiry. The bystander model can be set aside in favor of a powerful not-knowing. "Seen in this light, positions are expressions of knowledge, rather than structures that limit it." (272) This is free and open inquiry, "random" inquiry with no knower and no method. "It is not a question of learning new techniques or practices, but of giving up unquestioning loyalty to the techniques and practices that we are *already practicing* in every moment." (274) There is no target for fear and no position for gravity to crush. "Turning experience 'on its edge', it finds the meeting place of time and space, and there creates an opening for luminous knowing." (274) "Analysis turns from what is known to the 'coming to be' of what is known." (275) The dynamic of time becomes newly accessible.

Chapter 34 279–85

Rain of Questions K3

Questions can seem threatening, but any such response can itself be questioned. Emotional reactions are no stopping point; lack of energy is no excuse. When such questioning takes hold, "the familiar at once seems profoundly foreign."(282) Then we can allow knowledge to disclose itself "in the very act of analysis." We can investigate the limit-structure itself; investigate how our knowledge is put together. The free flow of questioning awakens inspiration.

Chapter 35 287–95

Knowledge in Human Being K3

Theory will close off knowing, but within experience, insight is available. Any particulars can serve the purpose: the way things begin and change; the history of our concepts; the inner structure of the things and relationships we rely on; the polarity of 'exists' and 'not-exists'. We discover knowledge to be available here and now, and do not have to freeze our knowings. This analysis can proceed in stages. Loosening our dependence on emotions and the pull of desire, we see them as energies; at a later stage, we see the energies as active patternings of knowledge.

Chapter 36 297–303

Magical Interplay K3

Knowingness can appear within what we do not know. The starting point is the knowing activity, the 'how' of our knowing. Instead of explaining, we look to the side, to the edges, and rely on self-reflection. "The mind engaged in thinking thinks about thinking." (300) Analysis "retraces through its own momentum the dynamic that experience embodies." (301) The knowledge gap disappears, and we find the knowledge 'between', integral to being.

Chapter 37 305–11

Knowledge through Wonder K3

Conventional knowledge responds to the problem at hand, and questioning readily shuts down. But if inquiry is grounded in the love of knowledge, we can question questioning itself, and in this way move toward intimacy. "The *attitudes* we adopt in carrying out our investigation shape the *attributes* we find in the world we investigate." (308) This interplay is interpreted in terms of subject and object, but it could be seen differently. Analysis is more than separation; it also depends on embracing and embodying, on appreciation and intimacy.

Chapter 38 313–17

Commitment Without Commitment K3

If positions are positionings, there is nothing for the self to give up, and knowledge can freely manifest. Positionings can express knowingness. We can know with the knowing of the logos and appreciate limits and distinctions as moves made by knowledge itself. Read-outs are self-founding, and the time they establish may not be the only time.

Chapter 39 321–27

Encompassing Opposition S3/K1

Conventional knowledge depends on opposition, linked to the polarity of self and world. But opposites are also related. We choose one partner in the pair over another, and although the choice is in a sense illusory, in this way we build up a world, filled with solid objects known by a self. The best way to challenge this is to look into the minding that chooses.

Chapter 40 329–35

Nothing to Know S3/K3

Based on opposition, distinctions have no substance. That is why they must depend on the momentum of the narrative for substance. But in another sense, they depend on the power of negation: the cutting quality of the 'not', which is more fundamental than the 'nothing' of non-existence. Negation is usually ignored, but it embodies knowing. The 'nothing' that negation knows cannot be referred to a something, or even to 'everything'; cannot be referred to a logos. Since the knower exists, it cannot know what does not.

Chapter 41 337–41

Nothing to Know K3/S3

Negation's 'knowing no distinctions' allows and accommodates, and the logos falls away. In silence, the activity of structuring beneath all structures becomes available. The 'nothing' of negation "restores to appearance the hidden dimension of being 'not established'." (340) There is no change, but transformation becomes an immediate reality.

Chapter 42 343–49

'Emptiness' of Emptiness K3/S3

'Empty' and 'emptiness' are introduced as a way of making negation acceptable. This risks establishing the 'ness' of emptiness, and thus negating the significance of negation. The 'ness' of 'emptiness' could be understood as unique: the 'ness' of 'ness'. But it still affirms. "Ontologically, this sequence imposes a kind of eternalism . . . Epistemologically, it means there will remain an unknown 'beneath' each act of knowing. And 'existentially', it binds and confines our potential, so that our freedom is cut off." (345) Different ways of dealing with designation seem unable to prevent such commitment; nor does the power of symbolic expression wholly succeed. This is more than word games, because words corral our being. We fall victim to juxtaposed presuppositions; we remain caught in the web of what is. "We seem to be playing a game of such sophistication that all possible moves, including all possible ways of cheating, are themselves part of the game." (348)

Chapter 43 351–55

Protecting Against Loss K3

We remain bound to the 'ness' and to establishing. Calling analysis 'word games' only reinforces the hold of what the words proclaim. The difficulty is that knowledge is misunderstood: "Knowledge as active engagement is replaced by knowledge as conceptual structures and models." (352) The potential for new knowledge is seized by the self for its own ends. "Proud of who we are, we confine ourselves to being something less than we could be." (354)

Chapter 44 357–61

Assertion of the Subject K3

The 'ness' of emptiness is also our 'I-ness', insisting that it remains exempt from inquiry. This nonavailability for knowledge is in fact critical, for what is fundamental cannot by definition by analyzed. "If the 'I' could be known, it could not be the 'I'." (359) But it always remains open to us to look into the arising of this fundamental. If the 'I' is part of the read-out, it cannot bind us.

Chapter 45 363–69

Knowledge Pervading Being K3

Knowingness can bring light to the darkness of 'ness' by continuing analysis past the claims of being fundamental. If 'ness-ness' is negated from the beginning, it never comes to be. We return to the knowing of negation, and find that knowledge pervades being. This is third-level knowledge. The parable of the three brothers and the different ways of understanding 'read-outs'.

Chapter 46 371–77
Inexpressible Wonderment K3

New knowledge endows form with a spacious 'quality'. "We may be ready to see that our words and stories are not names for a fixed reality, but symbols of a deeper knowing." (372) This happens not by leaving lower knowledge behind, but by embracing it. This is the love of knowledge. When we arrive at this insight for ourselves, we see no limitations in our beliefs. We arrive at the freedom of no positions. Committed to the reality of our suffering, we may reject this insight, but even our rejection affirms knowledge. "Like a spotless mirror, knowledge can reflect any image." (376)

Chapter 47 379–85
Knowing Implicit Illumination K3

Not-knowing becomes the gateway to knowledge when we can thaw it into positioning. We can move toward the challenge of truly 'doing nothing'. We can challenge every view and limit; can look especially at language and at the self. Such knowledge "deals with this world, not some other; it unfolds within our own experience and reflections, not someone else's constructs." (384). Things are left as they are, but are not that way. "The witness witnesses no 'ness'." (384)

Chapter 48 387–93
Opening Time and Space K3

Deeper knowledge can also open time and space. As an entry point, consider the failed claim that what is presented here is a 'path' to knowledge. Space and time can be understood differently by a different knowledge, for time, space, and knowledge share an interplay of intimacy. But this understanding must operate at a level beyond words and concepts: it must take up time and space as unknown and available for inquiry.

Chapter 49 395–401
Dance in Time T2

An example of inquiry through knowledge: the marvelous dynamic of time: the tapestry of experience. Time's momentum discloses knowledge and opens to being. To see this, it is enough to open one point fully, letting the rhythm of experience find expression, supporting communication.

Chapter 50 403–411
Dynamic Intimate Awareness K3

Knowledge reflects established limits: the 'sameness' of the same. But the 'ness' of that sameness establishes nothing. Looked at with greater knowledge, conventional time and space, which establish sameness, disclose a lack of knowledge in operation. Even the flavor of certainty that experience carries can be challenged.

All such distinctions could be considered exhibitions of an underlying knowing, which suffuses all appearance. Such knowledge is not prior, nor does it operate. What can we say of it? "First, it would be free of assertions, for assertion stops the free flow of knowing. Second, it would be free of judgments, for judgments require positions. Third, it would be free of distinctions, for distinctions accept the truths of a more limited knowing." (408) It brings remarkable freedom, for "nothing is as it is, yet nothing is other than it is." (409) We can embody such a global knowledge in a Body of Knowledge. "By nature knowledge illuminates . . . Like a lover absorbed in the beauty of the beloved, we can give ourselves to knowledge. The depth of our love will disclose the range of our vision." (410)

Knowledge of Time and Space

Chapters 1–3 3–18

Three Times T2

The relation of the three times and the source of time's flow raise a variety of questions; compare the critique in LOK. The past cannot be contacted; the future cannot be determined. It may make more sense to view the source of time's dynamic as being the self, active in the present. The notion that the present must come from somewhere—must have a 'source from'—may be a projection, an inference based on the present sense of past.

The present has a special intimacy that past and future lack. But perhaps the future gives present form, while the past gives it substance. We can speculate: perhaps past and future are linked to the present 'beneath the surface' in one great body of time; perhaps at their furthest limits, past and future, neither of which we can know directly, meet. Our view of time as one-dimensional seems inadequate, given the endless reach of space and range of knowledge.

We make assumptions about how time works; for instance, that it unfolds uniformly whether human beings are there to observe it or not. These assumptions are based on knowledge that arises *within* the framework of time. It assumes a certain model in operation, and while there could certainly be other models, it seems all possible models would exhibit the feature of continuity. "A fundamental rhythm . . . situates and positions. Whatever distinctions we maintain, specific 'dynamics' and 'mechanics' determine an underlying structural unity." (14) A similar claim of continuity is made with regard to the self, and again certain dynamics and mechanisms—including a 'from-to' structure—operate. The existence of the self may be a fundamental mechanism, but it is not that difficult to imagine entirely different mechanisms; for instance, a view of time in which all the events of an aeon merge into one moment. Does this last alternative suggest that change as mechanism is not fundamental to time?

Chapters 4–6 19–30

Experience in Time T2/T1

The from-to order of measured-out time is lifeless. The bystander is its guardian, applying the logic of the logos. At the center of the structures imposed in this way stands experience. Shaped by the concerns of the self, experience tends to intensify and to generate tension, as well as reasons, excuses, etc. The order imposed is superficial. Emotionality surfaces, destabilizing the painfully created first-level order toward greater identification. Life takes on oppressive and limited qualities. A surface fascination conceals sameness: A basic ignorance operates. The fundamental psychology of the self is generated, based on involvement and identification. Placed at the center, generating standard first-level read-outs of the logos, the self is cut off from the body of time. But what if time were at the center?

The self as owner is outsider to the temporal dynamic, and so are the objects the self knows. Everything is cut off from time's dynamic. Yet the self is always active, with an activity we could understand as as interpretations of time's momentum. From this different perspective, experience already draws on time's dynamic. It is just that the self appropriates this dynamic as the substance of its own being.

Variations in time's flow are seen as purely subjective, and we respond by trying control time and make it uniform. Fundamentally, however, time is against us, and it slips away. Perhaps by expanding consciousness, we could contact time more directly, but linear time chokes that possibility off.

Chapters 7–9 31–47

Linear Time T2

Linear time is an aspect of experience, arising as the projection of a particular awareness linked to the bystander. Experience is filtered to make sure it conforms to the prevailing 'order', "Linear time is the 'carrier' for experience, assuring through its specific 'carrying capacity' that experience will be experienced in a particular way." (32) Appearance exhibits to mind, which is also a construct communicating a particular message that confirms what we take to be real. Mind is the victim of the rules of the order; on the other hand, mind has some capacity to play with experience; it can also act 'into' time, shaping what happens. This suggests the possibility of a more direct awareness, linked differently to time, participating differently. As an example: the senses and sense intensity.

A temporal order arises through 'taking' time's momentum in a certain way that serves to link the inner and outer. This way of looking shifts the emphasis away from subject and object, which are now seen as patternings presented by the logos. Linear temporality is another such patterning. Underlying this is the second-level dynamic *becoming* of time, which we can investigate only as activity. In this activity, we find the bystander renewing its occupancy from moment to moment, in the context of some underlying persisting structure. In this way, time's dynamic is stripped of any creative potential. To break out of this, the best place to look seems to be the activity of the self in projecting self and world. If such looking becomes possible, the structures the self projects lose their solidity, becoming "inspired presentations" of the logos. (38) This second-level way of experiencing gives a momentum of 'just interactions'. This dynamic is not ordinarily available to the self; instead, the self experiences time as a hostile outside force. But this in turn results from a choice the self makes: "The temporal priority of the objective realm is based on a prior accepting of the existential priority of the self." (40) That is, the objective realm arises based on the self's need to exist. Other interpretive structures, and also emotional responses, derive from this fundamental structure: "attempts by conventional knowledge to work out the inner structure of the prevailing logos." (41) Bystander-self and the frozen momentum of first-level time prove to be intimately related.

Time flows; existence is an activity. "The classification into past, present, and future is a very rough approximation of this fluid motion," this gathering momentum. (45) Time seems less a device for measuring than a fundamental force. This force is channeled into an order, and reality is established. The dynamic remains in play, but most possibilities have been foreclosed.

Chapters 10–12 48–61

Time of Knowing T2/T1/K1/K2

The subjective and objective order follow two different times, like two clocks. The 'real' must conform to both clocks. But in the end, subjective and objective are complementary. They trace to the same order, founded by the logos. A temporal order is in play, allowing only certain interpretations, but the structures of that temporality can be challenged and investigated. Normally this is not possible, because we rely on language, which embodies the momentum of the temporal order. Yet language can also open to deeper knowledge; it can measure out, but it can also create. We can stay open to the deeper levels of communication that language allows.

Time on the first-level seems to control knowledge (for instance, in the temporal priority of not-knowing). Knowledge is steered toward preestablished positions. We do not know time itself, nor do we know with the knowing of the logos. But in second-level time, appreciation for the presentations of time is itself knowledge: Knowledge can know time's dynamic.

"[Second-level] rhythms of becoming allow for fluctuation between something and nothing, inviting a knowing that goes deeper than polarity and contradiction." (59) For instance, we could inquire into history: its cycles and rhythms; the ways it supports and does not support knowledge. Such first-level inquiry could make the second-level dynamic of time more available. Different ways of knowing provide shifting feedback that generate different rhythms, and this too makes time differently available.

Chapters 13–15 62–76

Measuring Out T2/K2

First-level knowledge and time are at odds, but second-level knowledge and time coordinate directly, through the momentum of the logos. Knowledge is seen to be given by time's momentum. First-level momentum and time are a particular instance of this. The image of the robe and its pattern: When intimacy becomes complete, the outcome is Great Time, or the Body of Knowledge.

First-level order is given as a read-out by second-level momentum. But this does not mean that first-level time captures anything "fundamentally accurate" about second-level time; in fact, the insistence on substance virtually guarantees misinterpretation. First-level existence is built up out of what second-level time presents as shadows or echoes: the residues of past experience. The apparent harmony between the temporal order and what it discloses thus reveals a deeper disharmony inherent in first-level structures. To get at this, consider first-level objects and events as the displays of an underlying temporal rhythm: specifications that are then regarded as points, yielding measurement. Even though it appears static, the measurement grid of first-level time—and the first level objects it discloses—express time's second-level dynamic. On analogy to the Heisenberg Uncertainty Principle, in determining the *position* of outsiders, we lose sight of their *velocity*. "Second-level time . . . teaches us to look for knowledge in transition." (71)

The dynamic of second level time offers rhythm without specification. This is not simply subjective. We reach that conclusion when we insist on substance, subjecting feedback to the pull of gravity, with indensification the outcome. But feedback can be left more open, making time available to knowledge, but this does not happen: the dynamic of time is measured out in such a way that it supports the interpretation of solidification. Such measuring out is accomplished through a certain kind of knowledge. This knowledge is limited, as seen by the observation that between any two points it specifies, other points could be specified, approaching the zero-point as limit. But then, if the limit is the zero-point, does momentum originate in the zero point? Momentum seems to originate with time, but how can this be, if time itself is momentum?

Chapters 16–18 77–87
Dynamic Momentum T2

"A focus on momentum lets us consider subject and object alike as projections of the underlying energy of second-level time." (77) Second-level time presents the logos that informs the first-level temporal order. Instead of subject and object, there are interactions, built up through and out of the momentum of time. This momentum is accessible as creation; we can be intimate with it. And even within the most frozen order, this momentum remains available.

Second-level time makes creativity available, for it is more fundamental than the imposed structure that shuts it out. In turn, creativity makes time differently accessible.

Time's rhythm engages a movement without movement, like floating. As it enters the order, it becomes points, origins, and boundaries. The commitment to substance allows for no gaps. Form takes form: "What *is* emerges as the echo of itself." (84) Points are taken as basic, but in fact rhythm precedes them: Points are superimposed in an "instant immediate emerging." (85) "The whole of appearance can be recast as simply this: An exhibition rhythmically transforms." (86) In the taking form of the order, friction arises, an expression of the demand to exist. A balance specific to the order emerges, yielding the rules of measurement and the logos.

Chapters 19–21 88–104

Zero Emerging Rhythm T3/S3/K3

Before the juxtapositions of the order, we approach the zero-point—the free dimension of zero—as the character of unconditioned existence. Here is the 'prior' of presupposition: a pure space. Here Being takes occupancy, in a merging in intimacy of Space and Time.

The zero is there in each act of observation. This is Great Time, where even the tenuous connection to form drops away, for form is the exhibition of zero. Lower-level events in time echo this zero. "If we are willing to abandon the structures of conventional understanding, we could describe this 'feedback'—the expression of the echo—as the origin of existence." (91) The challenge is to open phenomena to unlock their inherent dynamic as the echo of zero. The zero remains as it is; gravity cannot bend it. Without departing zero, time's rhythm emerges, not requiring multiple points. It offers "the flow of a luminous energy independent of motion . . . like the glow of life in a living being at rest." (92)

We treat movement as inevitably measured out, but that is not so. In silence, there is the movement of the Body of Time. All points participate in this "rhythm immediacy." In accord with time's rhythms, 'isness' emerges; seeing this supports the experience of intimacy, a touching with intelligence the free body of Space. Zero presents the rhythms of Time as the exercise of Knowledge in Space. "From this perspective, substance could be understood as a kind of 'holding pattern' against the force of momentum, a way of using momentum to support what persists." (97)

Great Time restores the limitless and open, cutting through disharmony. "The presentations of time 'explode' into magical expressions." (99) This is no mystical vision, no 'other' way of seeing. It is the circle of zero become the triangle of form, exerting the force of gravity. A momentum arises as a 'to' that points, giving rise to the line. "Zero is the . . . nonexistent totality of what would wish to be." (102) It is the rhythm of time excavating space, a projection of zero gravity.

Chapters 22–24 105–16

Invariable Time T3

Why does the first-level event arrive? The second-level answer is that time itself moves with its own momentum. But in a third-level view, time is not linked to momentum, which can better be seen as a mechanism through which an invariable time presents. A first-level perspective insists that what is seen is real; a second-level perspective reveals the 'order' that gives all appearance. The third level avails itself instead of the freedom of no positions.

Points and positions can vanish into the 'non-rhythm' of an infinitely open invariable. Then momentum vanishes. Temporal rhythm transforms into the knowing of the Body of Knowledge, taking form in the mechanism of knowing. Instead of presenting to present mind, Time presents an invitation to knowledge. This becomes possible when consciousness embodies knowledge fully.

Mind looking at time reflects back only configurations of past, present, and future. It makes no sense to try to "arrive" at a different understanding, because we are bound by the field patternings. Yet knowledge can work a kind of alchemy of time. The best move is to remain on the path of inquiry that understands the temporal order. One way to look: The three times merge in a hidden temporal dimension, where a dynamic without movement or momentum is available. A strategy suggests itself: allow the three times to merge and integrate. "If we could know just one moment of time in all its fullness, what would be left unknown?" (116)

Chapters 25–28 119–36

Opening Space S1

Space allows all appearance. Essential to appearance, silent partner to each object, it is as important as what appears. We ignore space, sure it has nothing to offer us, yet space has a mysterious power shaped by certain rules (e.g., exclusive occupation and unitary occupation). If space could be occupied differently, time and knowledge would be different as well.

Space in relation to objects is like the 'not' in relation to the identity of what is. While we ignore the not, we could challenge this ignoring by looking at edges until the notion of identity becomes deeply questionable. At the other extreme, the notion of space as non-existent is deeply problematic. More generally still, there is the question of the relation between space and what it contains. Space seems somehow active and present, without *being* anything. The scientific view is not inconsistent. Taking this further, one could speculate: objects are a projection of space. This speculation seems only of theoretical interest, with no way to embody it. But we can question this ready conclusion. What would it be like to engage matter as inherently spacious? Most likely it would be more open, allowing more freedom. Hence the value of a more open formulation: "When the features of the known world are not, this is space." (129)

Suppose we saw shape and form as blueprints that structure space, allowing us to communicate with it. The result would be a very different perspective on such phenomena as distance and existence, on the relation between space and time, and on such possibilities as multiple spaces and a space prior to appearance.

Exercise: Explore the relation of matter and space in the cosmos; of space and sky; of solid objects and their interior space; the interaction of 'layers' of space with the objects they 'contain'.

Chapters 29–32 137–57

Body Space S1/S2

Whether occupied or not, space continues on. But what space do we really occupy? In our lived experience, it seems we always occupy the 'same' space, carrying it along with us. It is our space: We own it. But what happens to it when we die? Does it merge into a more encompassing space?

How does the self enter space? If space must be physical to allow 'things', must it be 'intelligent' to allow a self that knows? When the self does occupy space, has space changed its character or qualities? Such questions render uncertain the relation of space to identity. They suggest we could perhaps experience space directly, just as we experience awareness directly. Then how are awareness and space related?. Is 'my' experience linked to 'my' space? Are the claims of the self, which structure all experience, linked in some way to space? Could knowledge be more space-like?

We can identify two levels of space: the space that co-exists with what appears (e.g., physical space), and the space that allows appearance (including co-existent space). Second-level space would structure reality on a deeper level than the structures we usually accept as basic, for it would be prior to all interpretive structures. Second-level space would impart transparency and luminosity to appearance, allowing for more knowing. The space/objects field could be understood as a projection of this second-level space, which *does* the projecting. If this is so, then all first-level appearance remains second-level space.

Exercise: Reflect on the way space allows an endless variety of properties, while having no properties itself. Even if it cannot be measured, space can be experienced in its allowing capacity, while each appearing property would reveal something about the space that allows it.

Chapters 33–36 158–76

Allowing Mind S1/S2/K2/T2

There are grounds for assigning an experiential component to space, at least by analogy. For example, "separation" has a psychological aspect. At a deeper level, there are rules for 'inner' space that seem to parallel 'outer' space. This may all be metaphor, but metaphor carries its own truth. In any case, our experience of physical space is also metaphoric, for what we know is air, not space. By treating the mental as available to space, we at least do not foreclose the issue in advance. After all, what appears 'in' space appears 'to' mind. If this view allows new ways of experiencing, we stand to benefit greatly.

Mental phenomena seem in a rough sense to have a physical location. They also depend on the space-like characteristics of boundaries and partitions, and on the allowingness of a mental realm. At the same time, the ways in which such mental space differs from physical space is also instructive; e.g., the fluidity allowed by dream space. Finally, physical space seems to depend on mental labels, even though we tend, perhaps invalidly, to dismiss such dependency as purely subjective. Even when we refer to an objective space independent of our perception, we have made it mental (have 'perceived' it).

When we turn to space at the second level, we can question the relation between knowledge and space. Knowledge knows second-level space as the 'appearing from'; in particular it knows first-level knowledge as the projection of second-level space. At this second level, the mental fits more easily into first-level space; indeed, *must* be a part of it. This makes first-level knowing insubstantial, giving appearance a magical glow or shimmer. A focal setting is set to work, an angle is carved out. The observer arises together with what appears and with the gap that separates the two. The knowledge gap between subject and object is the local representative of this more fundamental gap. Investigating it, we discover that knowledge can enter space, even if the temporal order sees to it that it appears only as nonavailable.

> *Exercise*: Observe the knowing of mind by mind, fully sensitive to the sense that what you are attempting may seem impossible. Let the frustration be a sign of what we do not know of mind. Look at how mind and matter are linked, letting this investigation lead to a more neutral zero-space 'before' mind and matter, the space where distinctions arise: "presentation before possession or position."

Chapters 37–40 177–89

Space Field S2/K2

The knower emerges out of a background, situated in a place, acting as bystander. This locatedness, conditioned by space, will also determine the nature of the space that investigation discloses. But it is possible to challenge that space by considering it as the "shifting exhibition of a field." (178) Within the field, mental and physical unfold in tandem. First-level space draws from the field the ability to allow appearance to appear. Even 'given together' is an aspect of the field. But an entirely different field could operate. Still the field formulation remains tied to first-level entitative understanding, and in this sense is transitional. (cf. 204)

When the field is not seen as such, characteristic limits on knowledge result; e.g., the mystery of the power of space to allow, or the barriers between mental and physical. The recourse of exploring field dynamics and mechanics is not available. For example, only what exists can come into being. A gravity sustains a mechanism. But because the field is comprehensive, if gravity gave way, the change would be dramatic. Perhaps such changes are fairly common, occurring without our taking notice due to restrictions in the prevailing field.

An innately indeterminate allowing may be a part of each field. Like first-level space in comparison to first-level objects, the field may not be as rigid as what it allows. Second-level knowing, attuned to the field dynamic, could know the field as a projection of second-level space. Even in first-level terms, the field is somewhat fluid; e.g., the realm of sense experience under a microscope, or the arising of new insights. Looking at border phenomena can be an effective method for exploring the field. Appreciation for the space-allowing that gives field exhibitions is a vital starting point.

> *Exercise*: Look to the edge, looking at space. Repeat this at different scales, looking for the 'properties' that space (as part of the field exhibition) manifests. Look especially toward the merging of the mental and the physical.

Chapters 41–44 190–210

Field Dynamics S2

Elements that occur together share a certain flavor or dynamic; in the same way, all appearance, existence, and experience share in an all-encompassing field. The field is self-establishing: A tentative order gives an incipient structure; feedback consolidates and structures further; a lineage takes form; embodiment manifests as the net outcome of field interactions. The field gives time, space, and knowledge. Because it encompasses, the field as such is not a part of what is given, but we discover the field in its dynamics.

Field dynamics allow certain mechanisms to operate. Each specific, shaped by these mechanics, implicates the field as a whole. Even the distinction between form and space, something and nothing, even the mind that accepts these distinctions, is a part of the field. Internal claims to authenticity cannot be binding on inquiry. To go beyond these, we can look at space as what allows, arriving at "the 'field of the possible' as the *negation of its impossibility*." (197) Obstacles are not 'not not-there'. Here the 'not not' expresses the power of allowing.

Compare an electrical current that flows when a gap is not there, and then consider 'not not not-thereness' as the potential for flow. The field encompasses negation.

The field dynamic allows for the mechanisms that give rise to existence by not preventing them. By focusing on 'not the field totality', we arrive at the field as "the active allowing (non-prevention) of all that does not 'not-exist'." The rhythm of the field dynamic determines what that dynamic will make room for. Interrelations among the elements both express and contribute to the dynamic. Such rhythms can be directly experienced. Characteristically the field of what exists exhibits polarity and the frustrations of desire, as well as an improvement orientation for a self. But the self can be challenged; for instance, through a focus on change and transition.

> *Exercise*: Look at how thoughts open into space, and then look into a place where thoughts themselves are space. Go into the feel of this space. We can see that space is alive, not bound to 'from' and 'to'.

Chapters 45–48 211–26

Senses in Space S2

The realms of the senses—including the special aliveness of the knowledge that knows what is sensed—reveal the magical ability of space to allow. To try to explain this ability only seeks to have what goes beyond the field conform to the field.

Explanation cuts off the rhythm through which knowledge can arise, given by the rhythms of the field. But as expressions of the field, even explanation and reasons manifest the field dynamic. One way to discover this is through appreciation of first-level space-time-knowledge interactions; e.g., the vastness of space, the flow of time, and the range of the knowable. These in turn manifest in the senses and their coordination.

Inner and outer space interact to produce knowledge, but this interaction must be based on an initial unity. Form finds expression through time, activated through the rhythm of the senses and perception. All this is labeled and made meaningful, but the underlying rhythms remain available despite the solidification and the tendency toward emotionality. "After all, when it rains, does space get wet?" (221) It is the arising of the senses, rather than the content they communicate, that seems most interesting, for here the dynamic of the originating rhythm is retained. We can contact it by taking the drama of our lives as play, as a source of enjoyment. We can play our role forward, and discover the active embodiment of the space blueprint.

> *Exercise*: Investigate the operation of the five senses, considered as a communication 'by' space. Extend this to the mental realm as well. Let the feel of experience be unknown. Open the entire field within which senses and mental events arise, including whatever interferes with sensing, such as thoughts. Relaxing into the feel of space helps the senses to open. A feel of vastness engages.

Chapters 49–52 227–240

Allowing Matrix S3

"Space is not a candidate or partisan; it takes no one's part . . . Never participating, never occupying, it never excludes. Only through space can 'gravity' work its way." (227) Within Space, there can be no gaps, though this is how space and form present themselves. Space is the infinitely allowing matrix, "the complete accommodation within which all points are inseparably one." (228) Each known point is a gateway to this allowing. We in turn can allow this openness through our vital questions. "If, in the course of this inquiry, the old assertions prove hollow, we can follow the 'hollow' into the allowing negation of Great Space." (229)

All appearance seems to require a surround. Even illusion must have space to appear. "Allowing is an exceptional gift, one that in turn gives everything else." (231) "Just as an exhibit in a museum exists both as an an exhibit and as the specific work it claims to be, so Great Space 'exhibits' appearance, both as an exhibition by space and as what it 'is'." (231) Its openness is linked to knowledge through negation. The temporal order makes of this negation an asserting, establishing rather than exhibiting. What is more, it does not even acknowledge this building up, insisting instead on the preexisting and on the self as its 'in-habitant'. But that is just another interpretation.

In describing space operation, we arrive at 'nothing to say'. But this 'nothing to say' is not a stopping point. Instead, it leads us into rootlessness, the unbounded, and the unstructured: a vision of completeness and perfect accommodation. It is an aliveness that does not depend on 'is'. There is no restriction because there is no becoming.

> *Exercise*: Trace the lineages of whatever appears, in any and all dimensions. Let substance go. This will invite a shift from first-level to second-level space, from space that contains to 'space' that projects. We discover this: The object of the object is also the object of space. To refine this further: "The object of the object is space. Existence is responsible for its own appearance."

Chapters 53–56 241–55

Great Space S3

Insisting on identity, we become strangers to Great Space, visitors taking advantage of its hospitality. We assign Space an identity as well, and thus create the possibility of using it up. But form is still space. "Space projects Space into Space, in an exhibition that ripples outward." (242) Subject and object arise as echoes, as two pyramids that meet at the apex. They are not the outcome of a causal sequence. The point singularity allows lower-space accommodation.

Each singular point is also the zero of being, punching a hole in the fabric of existence. Since this is so, the singularity does not establish; the point is preserved as available. Great Space punches the hole into the whole, without ever founding. It is when we punch a hole into the hole [a whole into the whole?] that founding occurs. We conduct space toward the physical, and toward temporality and 'mine'. But we can do it differently, balancing in the zero point. We can establish the known world, but we cannot deny what is spacious in space.

When Knowledge embodies space fully, it becomes the Body of Knowledge. "Appearance exhibits the quality of light." Space itself is alive, the exhibition of what is. Knowledge appears as beauty. "Like two lovers exchanging kisses, Space and Knowledge find each other delicious." (252) Through the senses, space and matter interact.

> *Exercise*: Look at the relation of subject and object and reverse them. Now look at the relation of space to existence; start if necessary with thought and awareness. What determines the boundaries of the subject? How is it related to space? Is bystander subject or object? Or is it space 'as' subject? Can self be understand as the allowing one? Does space appear as well in the appearance? As space communicates space, it plays the role of both subject and object, and thus plays no role at all. It is non-originating source.

Chapters 57–60 259–76

Before Knowledge K1/K2/K3

We start from ignorance, so "gathering and implementing knowledge could be understood as the fundamental human activity." (259) But in the background, the unknown remains primary. From a knowledge perspective matters look different: Knowledge tends by its nature to know, so why should there be 'not knowing' at all? Why is the infinite range of the knowable not matched with a boundless knowing? The question seems to hypostatize knowledge, but rejecting the question condemns us to not-knowing. The difficulty may be that we do not know mind. This is the fundamental not knowing. Its consequences are that problem and solution stand at opposite poles, repelling one another, and we "must expend prodigious effort to bring them together." (261) And even then, the communication is incomplete, and the problem emerges in another form. But problems can also be understood differently, as the read-out of a more global knowing. Could we then gain access to *that* knowing? If so, even our repetitions would be seen as creative. All we need do is see the well-worn patterns at an angle. Then we can cut through the momentum of sameness. Looking at boundaries and limits, we can turn to not-knowing as our teacher. "Without risking either laziness or craziness, not-knowing teaches us to see." (263–64)

Not-knowing might represent 'x': the potential for a new knowing. X is unknown, but also infinitely variable; x also marks the spot of buried treasure. In lower-level knowledge, x is the unknown field from which knowledge emerges. At a second stage, a focus on the limits of the known opens this field up. At a third stage, x is the potential for new knowing, the carrier for knowledge; it is even the active knowing within first-level conventional knowledge. At the fourth stage, the unknown is the sponsor for knowledge. "Newly familiar with not-knowing, we see the way in which we requisition a previous set of arrangements from the storehouse of what is familiar." (268) Not knowing "now delivers knowledge in the non-deliverability of knowledge." (268) It discloses the x of the axiom, and opens a second-level knowledgeability.

We relate to knowledge as though it were a possession, and thus bring not-knowing to the fore. But there is a prior point, a decisive point, a zero point. A focus on content looks past the zero, but zero is the point of active knowing, the point of not-knowing. We can rediscover it as the zero-point within, finding within it the body of knowledge. "By standing in contrast to what is, 'zero' grants existence permission to be." (271) (Compare zero as placeholder.) The unknown zero lets us measure out a world. It is like discovering a black hole at the center of every atom.

We turn from the zero point to pursue our concerns, but turning away prevents knowledge from manifesting. In theory, we could let knowledge be, without regard for concern, bearing witness to itself. But why would we do this? In asking, we plunge back into the world of concern. Then does 'concern' itself express knowledge? This would be an intrinsic knowledgeability: Great Knowledge as the not-truth of positioned knowledge. "Through our exploration, the *gap* between known and unknown . . . can become the *root* of the unitary event of active knowledge." (276)

Chapters 61–64 277–96

Path of Inquiry K3/K2/K1

Inquiry can aim at Great Knowledge without heading anywhere in particular. Even though we may wish to make great efforts, being—as it is—is knowledge. Great Knowledge is found not as an outcome, but as a gesture made by Knowledge itself. It manifests in acceptance, balance, and appreciation, especially for history and for the patterns and limits we weave.

To claim "I know," savoring insight and discovery, risks saying, "I stop here." We need a way to say "I know" that keeps inquiry open; that sees knowing as process; that recognizes the limits of "I know" and especially its relation to the known. "One point can proceed along different lines, in accord with varying momentums, subject to a range of responses, giving multiple meanings." (286)

The quality of our knowledge proves itself in the quality of our lives. The established order fills us with care, yet we have no way to escape. It is like being trapped in the narrow angle of a triangle. But if we can embrace polarities, without commitment to the order, we can let inquiry move us toward knowingness, as aspiration deepens into love.

Knowledge makes the known order unknown, revealing the order itself as knowledge. We can follow the path of the professional, making knowledge our vocation, dedicating ourselves to knowledge. Discovering the patterns within experience, we recognize them as the manifestation of knowledge. Now we are free to join the lineage of the great knowledge holders, join in a partnership of time, space, and knowledge: a correlating co-appearance.

Chapters 65–68 297–317

Self Center K1/K2/K3

Our ordinary human knowledge has intrinsic limitations, based in its origins as a device for the self to deal with the world. It unfolds via rules based on the subject/object structure and a corresponding space and time. We look to sequences and consequences, like someone tracking an animal's footprints [or the shadows on a wall]. We leave certain questions out of account. Rhythms that do not allow for the operation of the self and self-perspective are in turn not allowed. Four chief conditions operate: specifiability, sequencing, transportability, and congruence. The resulting limits are not even seen as limits; they are inescapable aspects of the governing order.

The lower-level knowledge of the self depends on the senses, and thus on repetition and familiarity. Still, knowledge of the self itself must arise differently, for the self is the unique being at the center of experience, For instance, to look for a physical explanation of self presupposes the distinction between subject and object we hope to answer. To accept a separate mental realm raises the issue of interaction. An idealist view raises its own difficulties. The evidence suggests all these formulations are flawed, mistaken.

In the bystander model of knowledge, the distance that separates knower and known is problematic. The bystander occupies a wholly different realm, separated from the known by an unbridgeable gap. The bystander, though here, proves unobservable: the most distant of all. On the other hand, we know very well the 'feel' of living in a bystander world, situating what arises in a series of momentary knowings. The bystander becomes a bias-stander, and discovers a world in disarray. A crucial dimension seems hidden. That dimension is what we want to discover, by understanding the bystander.

The bystander operates within a field, which specifies the awareness in operation and the forms it can take or open into: sense experience, emotions, etc. If there were no bystander, each such form would exhibit awareness directly. Here may be the hidden dimension of knowledge: awareness active *within* the field. We can turn from the self to investigate the world as a specific presentation and exhibition; we can turn from the content of what presents itself to look at the interaction of space, time, knowledge. Then we ourselves become the topic of inquiry, see knowledge as the pervasive medium, and know the read-out as read-out.

Chapters 69–72 318–34

Body of Knowledge K3

In sheer space, conventional distinctions of lower knowledge vanish. Is floating helplessly our only alternative to the gravity of positions? We do not have sense experience, nor do we have logic or reasoning, which ultimately trace back to the senses as well. Once there is more than one, we move toward multiplicity, and the measured-out world takes form. But we can turn again to the unknown, the 'x' within each axiom.

Shape and form do not give the whole. There is also the 'reality' of energy and the vitality of mind in operation. A multiple dimensionality operates, interacting with the unity of being. Given this unity, where are the limits? We can see the openness of knowledge before it is measured out. We can cultivate attitudes that support knowledge. We can resolve to embody knowledge, to unite with the Body of Knowledge. Attuned to the feel of this possibility, "we can leave the open open." (325) We can cultivate inquiry, asking, "How does it feel to be kissed by space?"

Mind is the guardian of conventional knowledge, safeguarding by establishing routines. But all this is circumscribed by our sense of our own needs. If we drop the focus on security, we can look at mind itself, discovering in each obstacle to knowledge an expression of knowledge; discovering knowledge as capacity. Experience is available for alchemical transformation toward knowingness. "Within all 'minding', the Body of Knowledge offers full, illuminated awareness." (329)

Mind allows for multiple interactions, but perhaps all this is simply the product of specification. At a deeper level, experience identifies, assuring that we turn at once to the past. Mind gives us what we care about, but why should we care about caring? "If what mattered changed, knowledge and experience might not be bound in the way we imagine." (332) The whole of being can be inseparable from knowledge. In its interweaving, even contradiction may not matter. There can be no measurement, nothing, to point out, and nowhere to go. "And if I do go, how can I distinguish the 'I' that goes from the rhythm of going?"

Chapters 73–76 335–51

Creative Not-Knowing K1/K3

Any starting point—e.g., substance—can be the starting point for inquiry. At some point, though, we reach a stopping point, a 'nothing to say'. That 'nothing to say' underlies each saying or knowing, as shown in mistakes, frustration, etc. But mistakes could be points of entry for an expanded knowledge. They also help us see first-level knowing in a precise way: as a proliferation of positions, each grounded in not-knowing, so that knowledge becomes almost inaccessible.

Not-knowing manifests in the restrictions on what is known and *can* be known. As long as knowledge is confined to taking positions, such boundaries remain intact. There is known and not-known, in a pattern of dichotomy characteristic of systems of knowledge. This leads us to the 'not' of negation. But each system refers only to itself, and establishes nothing. In the end, the witness proclaims not-knowing rather than knowing, proclaims that positions are unfounded.

The limitation suggested by the unfoundedness of knowledge also has no witness. Or could there be a non-positioned witness? There have been attempts to find one: objective matter, an unobtainable absolute. etc. But ultimately these all fail. Then could the unknown be the witness for knowledge? This returns to negation, which is also not established.

The 'not' of negation ('not A') seems to depend on a further 'not': A as 'not not A'. Here there seems no concealed affirmation, no hidden positioning: a double negation does not establish. The vehicle for such perception is realization—a realizing that allows 'is' and 'is not' to arise as a whole. This is a way to go past opposition. From shape and form, we go to the senses, from the senses to speech. From speech we go to meaning and value, activating the Body of Knowledge.

Chapters 77–80 352–367

Name and Form K2/K3

Before the prevailing order, we can speculate, there must have been a consciousness without name and form. But then universals arose, prior in a sense to specifics and the singularities on which aesthetics focus: 'blueness' rather than blue. Before name and form, there would have been singularity. In this singularity, we could stay at the level of zero—the space of thought before the move into the 'positive' dimension of cognition. This 'mind at the zero level' may still be accessible, if we drop positioning. We can arrive at nameless knowing.

Zero may also be available in transitions, when old names break down. Our present time is rich in such transitions, though the main indication is difficulty and confusion. Perhaps the naming activity itself will change.

If name and form themselves changed their operation, as might happen in this time of rapid change, would our ways of knowing alter? The potential for change seems real, but we have no direct way to access it. Instead, "we can explore the momentum of naming and giving form that shapes our current reality." (362) We must ask questions that we then allow to become active in our minds. As a natural corollary, we find ourselves questioning time.

Names express a certain rhythm of knowing, generating concepts, points, and meaning. "Language conducts perception toward reality." (365) On the other hand, language itself is an artifact, generated out of what has come before. If a truly unique reality unfolded, it would be inaccessible to name and form. Name and form give the real, but do so without establishing. "The known gives itself, happens of itself." (366) This giving could also happen differently; e.g., in art. The rhythms of knowledge could manifest differently. The giving by name and form is a hidden conducting of the dynamic intrinsic to all points.

Chapters 81–84 368–83

Shaping the Known K2

Perception depends on previous concepts and labels, based in the end on distinction-making and negation. We go from yes to no and no to yes. Based on agreement, the truth of what is perceived is established. Gradually the direct perception recedes. But it is not enough to return to direct perception, which is likewise based on shared understanding. Instead, we can go inside the label that has been set up, opening a new dimension, or going off at right angles to arrive at multiple dimensionality. Reality becomes determinations arrived at "through the transitional operation of indeterminate 'fields of knowledge'." (371)

In a field of knowledge, categories interact, yielding a final act of recognition and reaction. Stages move from one to the next on the basis of being interrelated, yet each also has its own field and feel. The challenge point comes in looking at the transitions between fields; e.g, the communicating forward of meaningful distinctions. For each of the senses, a specific field mechanism operates.

Every field of perception gains its specific feel through certain rhythms and relationships. Output based on feedback looks to past fields, then reads out the input to establish it. A focus on the field itself breaks this momentum. We can go beneath awareness and consciousness, themselves the outcome of a field. This questioning can be applied to time and space and knowledge themselves. In each case, a readiness to respond is processed toward a specific field. But even that readiness, the initial awareness, is field-determined. Thus awareness and experience must enter into partnership. Seeing this, awareness and also knowledge need not be confined to fields in the same way. Standard rhythms that lead toward awareness of content can give way to a lucent knowing.

Just as the senses shape experience, so our mental dispositions—modalities of minding—shape experience in dependence on specific fields. All such distinctions, however, can be seen as interactions 'within' awareness, exhibiting greater degrees of aliveness. We can open up each feel, arriving in the end at knowingness. Seeing how the point is made, we arrive at the 'decisive point' in all creation.

Chapters 85–88 384–96

Knowledge Ability K2/K3

As a model for the arising of the knowable, we could say that feedback leads to recognition, followed by perception and pronouncing: a threefold structure of projection. Positions are born of symbols that are not positioned, originating in zero gravity. "A center of gravity emerges, yielding the central point: 'Here I am'." (385) A gravity of polarity yields instant proliferation and the possession of the position, excavated as the known. But in second-level terms, all this remains more open: Knowing arises when the knowable is possessed by knowingness.

By opening focal settings in space, we open shape and form, allowing more light to enter, infusing appearance. Distinctions lose their significance but not their distinctiveness. Similarly, we can access the energy manifested in time. Each object can arise as space and each event as time, with the separation between knower and known arising as knowledge. 'Coming into being' can be seen as an interplay of space with awareness, the outcome of a subtle 'wanting to know' that manifests in the first instance through the senses. The outward momentum of the senses establishes the existent. An "initial candidacy" is accorded accreditation, and symbols and meanings and language confirm its claims. But from another perspective, what appears remains the manifestation of light.

The senses establish, but the original intimacy remains available for inspection: the allowing of Time by Space, returning Knowledge to Being. The senses operate in accord with the order, but we can open the senses. "We do not need to cultivate 'inspiring' thoughts or 'special experiences': The path to knowledge depends on greater appreciation for the conventional realm." (393–94) This opening of the senses is a test case for the availability of knowledge.

"When the senses can embrace all-presenting time and all-allowing space, we hear the echo of silent sound and see the shadow of the formless." (395) Intimacy opens space, and benefits flow in a stream of benediction. There are nine progressive stages in this opening: tool to vehicle to channel; friend to agent to master (as friend); becoming the best of friends, giving intimacy, and finally the fulfillment of love. Intimacy becomes the source of experience, and we look to the light of awareness. Reaching out through the senses, we discover our own hearts.

Chapters 89–91 399–410

Determining Knowledge K3/K2

All knowledge arises within the Body of Knowledge, which we can explore through inquiry. "Inquiry unfolds into intimacy in a process that heals as it reveals." (400) The being of the one who explores is brought into play, together with all that appears. We go beyond all points. We can determine that we are the carriers of Great Knowledge and discover a knowing inseparable from being. The senses project the zero point of knowledge without projecting 'into'; they project interaction and intimacy.

We can discover the built-up first-level reality to be the outcome of the implementation of knowledge, 'measured out' and maintained in a pattern of feedback that establishes the fundamental mechanisms. Tracing the source of this system leads us to the patterns of perception, but then to the values we assign, which are fundamental. Values in turn lead us to the presupposed, through consensus and previous determinations. Behind this lies human identity and the patterns of identity in general, but they can be challenged through a focus on the field. Now we can look at the feel of the field instead of the content: we can discover a drama very different from the usual drama of the self. As the play unfolds of its own accord, we are led differently, toward the creativity of the mind. Exploring this play, we find a fundamental 'aliveness' of awareness and an unexpected ability to act 'into' time and space. [A very fine summary of many different themes.]

We think of experience as first arising and then presenting itself to awareness; as a result, we assign awareness a grasping quality. But if awareness arises together with experience, knowing need not take positions. Experience is no longer experience 'of' something, for it integrates space and time with knowledge, and nothing is outside space and time. There is no 'from'; instead, there is 'only knowing'. Thus knowledge manifests in the inquiry, not in the results. But because we take time and space the wrong way, we miss the simplicity and radiance of 'only knowing'.

Chapters 92–94 411–25

Knowledge of the Logos K2

What seems fixed is an aspect of the order, and the order is given by the logos. What is 'true' must conform to the logos. This is so in every domain, mental or physical. Beliefs, etc.

present themselves to the logos as candidates for adoption; some of the adoptees become absolute, in the sense of being presupposed in all acts of knowing. Conventional knowledge gives expression to the outcome, as a living being gives expression to genetic coding.

The knowledge of the logos is available in first-level terms to be known by a knower. This knowing is 'cognition', which measures out across an infinite range of knowables (though generally subordinated to the goals of the self.) Typically it leads to explanations that offer the prospect of control. Cognition is often limited and faulty, though it is possible to generate cognition that at least has the merit of conforming fully to the logos; e.g., through the self-correcting modes of inquiry of science. The more the focus is on the growth of knowledge, the more cognition will extend to the full range of the logos. But there is also another, more far reaching possibility, which is to know 'through' the logos, to embody the logos. Cognition can move toward such knowing when inquiry is open and free.

The logos determines what forms of cognition are allowed, and also what shall count as real. "The known world and its contents, including the self that knows both world and contents, are nothing other than the explicit 'fabrication' of the knowledge implicit in the 'logos' 'through' the knowing activity of the self." (419) But second-level knowledge can open up all such absolutes. In doing so, it discloses that "just as they are, without being 'reduced' to mental phenomena, the realities of conventional understanding 'are' knowledge." In cognition the logos is constantly being communicated, whether through knowing or not-knowing; in fact, not-knowing that arises as the fruit of inquiry is the gateway to second-level knowingness. It invites knowing *through* the logos, stripping first-level structures of their claim to significance. Not-knowing, in the form of mysteries and inconsistencies, is also the way to break through the set that sees the logos only as object. An example: the relation of subject and object. Seen in terms of feedback, the object understands itself; subject and object "are interlaced in a web of knowing, woven in the rhythms of the temporal order." (424) Looking in this way, we arrive at a knowledge that embodies the knowingness inherent in the situation as a whole: a knowledge that need not emerge from a state of not-knowing, but is naturally present.

Chapters 95–97 426–39

No Path K3

The fruit of inquiry is Great Time, Great Space, and Great Knowledge. But from a Great Knowledge perspective, there is no fruit. It is like looking at a pyramid from above: the apex is also the center, already available at the outset. Step-like structures grow out of a first-level view. Then how can we make the bridge to Great Knowledge? We might say: By opening the points within the field. But here again, we rely on a procedure. When positions are seen as positions, the point is already open. What holds at the first level has second-level counterparts but at the third level the whole picture breaks down. There is no restriction and has not been. "Pointing points out the dynamic of being." (430) Time, Space, and Knowledge are the path to their own realization. "The vision is in the seeing, and the seeing is in us." (431)

We come to see that there is no path gradually, as we let go of the residues of a view based on substance. A preliminary move in this direction would be to acknowledge an intimacy between subject and object.

Time, Space, and Knowledge are not absolutes, and if absolutes appear, we must be ready to question them. They may seem to operate as absolutes, but if this is so, they do so in a way that sets up no structure and has no outcome. Global knowledge, which knows 'through' the logos, is the expression of this insight. It makes it possible to accept and appreciate the prevailing order without making it binding. It shows that "good" is 'good for' and "bad" is 'bad for': that values and meanings depend on the order.

Chapters 98–100 440–51

Body of Knowledge K3

Every form of knowing establishes a body of knowledge. The body of knowledge has its limits, revealed at points of paradox and not-knowing. Usually these limits go unchallenged; we accept the role offered us. But if we can transmute limitations, we find the Body of Knowledge active and available. One way to touch it is through embodiment. If we embody in our being appreciation for the energy within all presentations, releasing into the openness of space and questioning the patterns of lower-level knowledge, we may arrive at the Body of Knowledge without limitations—a fully embodied knowingness.

"Out of the beauty of perfect expression comes the 'aliveness' of Knowledge." (446) Nothing takes form, nothing points 'toward'. "The Body of Knowledge is not a position body, an ownership body, a category body." (446) There is no 'to' or 'from', no derivation. Even beauty is a kind of sign; it becomes 'beautific'. Knowledge has no purpose, and so is free to command Time and Space. Time is the vitality of the Body of Knowledge, never confined. Knowingness remains fully available, even when lower-level interpretations are in effect. A lower-level indication of this is co-existence; e.g., the co-existence of existence with non-existence, or of existence with space.

Mind, form, their interaction, etc. are a unified lower-level display of the interplay of space and knowledge in time. In ordinary reality, this interplay appears as emptiness: intimacy as vacancy and oneness as negation. But intimacy remains in the form of appearing, seen in terms of coexisting. "When we discard even 'potential' and 'field', the invisible and boundless, the 'non-existent impossible' can be created as unique."

Chapters 101–103 452–64

Toward Time and Space K3/T3

Knowledge can operate without the true; can operate pragmatically and in terms of presentations in which the mind itself can join. Appreciation engages such participation, "shaping a symbol that points beyond what is given to the giving." All appearance becomes akin to art. The drama of existence unfolds like a well-loved epic. There is no becoming; there are no transitions. The incomprehensibility of the knowable does not conflict with a unitary knowledgeability. Even patterns that repeat endlessly express the creative surge. We can reunite, reemerge, and renew. Thoughts and forms and senses manifest as their own creation.

Inquiry as Great Knowledge touches the giving point of time as the Body of Time. "In each emerging instant there is a merging with the momentum of time." (457) The rhythm of time is mastered, giving Knowledge as stable, Time as invariable, and all presentations as Space. The patterns of the temporal order are seen as luminous.

Within invariable Time, points become symbols. Nothing is held back. There is sameness in difference. All angles become one point. In one sense there is nothing above us; in another, we have lost our way. But the Body of Knowledge offers the gift of light, revealing its own radiance.

Chapters 104–6 465–76

Inseparable Presentation K3

Whatever appears seems to depend on a background, transparent and so inaccessible. If this background becomes a basis, whatever it claims to establish becomes unknown. But knowledge as the background escapes this limitation. It presents its own luminous knowingness. There is no stepping outside what is or what is proclaimed; rather there is a stepping 'into' appearance as appearance. With no beliefs, we are free to act on knowledge directly.

Great Knowledge does not go beyond. If we understand this as a point describing Great Knowledge (e.g., that all points are open, or emerge from one point, or manifest light), we fall into first-level constructs *about* Great Knowledge. But if we deny such a point entirely, we make Great Knowledge into the final point, the stopping point for inquiry. Then how can lower knowledge and Great Knowledge, for all their inseparability, co-exist? They cannot, as long as we regard them as mutually limiting entities, given by linguistic convention. But all such distinctions can be stripped away.

Great Knowledge cannot be communicated, but it does not need to be. Communication in this sense is based on a model of transportation, and there is nothing to be transported. "Great Knowledge is active as the inner light illuminating all that is known." (475) It is the open momentum of presenting. We need not choose or take sides. Knowledgeability is available.

Chapters 107–9 477–89

Decisive Creation K3/T3

The patterns we live by form the record of the whole, whose tracings we patiently transcribe. We can trace the patterns of this whole, and in this way propose to know the whole. All this depends on the mind that knows, yet mind is itself an artifact. It is *knowledge* that traces and records, with subjective mind a part of the fabricated record. Lower knowledge closes off, confining what can be recorded, but that is not a problem. "As it closes off, potential invites Great Knowledge. As it specifies, Great Knowledge accepts the invitation." (479) Creation itself is the story of the whole, embodying in the actual as the 'point of decision', prior to all specifying distinctions. The point reads out its own appearance and enacts its own decision.

The realm of decisive creation is wholly open, even when all the important decisions have been made, established, and confirmed by the logos, for the display is self-arising. Knowledge manufactures magic. The magic may become logic, but the decisive point remains available, for the foreclosing of linear time does not affect it. Each decisive point gathers together the whole. "In this ongoing emergency, all that is is invariably at stake." (484) Freedom can always be recovered. Between the rhythm of time and the decision points of knowledge, a sustaining balance holds.

The singularity of the point need not be interpreted as the starting point. There is nothing to initiate and no need to point out or proceed point by point. Specification will lead to first-level outcomes—in psychology, epistemology, and ontology—and creativity will fall away. But within all this, each point remains zero and open. We have made our point, but we can unmake it also.

Chapters 110–12 490–99

Intimacy K3

Characters and qualities are just their own presentation. A field and order are always in operation, but we can question them. "Without reducing appearance, . . . we recover for Being its nature as light." We arrive at the zero point 'beneath' appearance, and see the Great Knowledge 'no difference' between zero and not-zero. We enter the fullness of 'nothing at all': the partnership of Great Time, Great Space, and Great Knowledge.

Inquiry depends on inspiration and aspiration and these in turn express knowingness: a way for the light of knowledge to shine through the patterns of our minding. The desire to benefit personally is one more barrier to be investigated.

We start with appreciation, which deepens into wordless wonder, awe, devotion. Compassion follows naturally, helping deactivate the limits that center the self and awakening the potential for transformation. The final inspiration comes through love. "The Love of Knowledge is the Love that Knowledge extends toward us." (497) It is the Body of Knowledge disclosing Being as Love's Body. We could call it Great Love: the unfolding of Knowledge itself.

Visions of Knowledge
Liberation of the Modern Mind

Introduction 55–66 K2

Current thinking assigns knowledge to either the practical or the spiritual domain. But this dichotomy is simply a read-out of current dynamics and mechanics. The aim is to investigate the read-out dynamic itself. Such a shift might prepare the way for true transformation.

"By imagining it could all be otherwise, we soon grow vitally interested in the ongoing emerging that is 'right now'." (57) Imaginative intimacy lets us teach ourselves. "We can simply engage immediately: every instant a present act." (57) The test is this: do we achieve our own well-being.

Concepts rely on 'is' and 'is not'. But we can draw on them more lightly, integrating them into direct experience, penetrating to the heart of thoughts. "When concepts open to imagination, the fruit is the special lightness of speculation." (59) Illumination enters appearance: We *live* light.

Why do we choose against knowledge? Perhaps we value our limitations more highly. We affirm our expectations at the price of accepting our limits. We insist on the right to determine in advance.

In advancing knowledge, we must learn to use language differently: to hold meaning without depending on structure. "We must evoke . . . a telling that remains a questioning and a questioning that is also an imagining." (61) The first TSK book tries to do this in one way, unfolding the presentation in such a way that no claims can be maintained. Language must literally become more sensitive: more attuned to the physical senses and the immediacy of experience. It must make available the extraordinary within the ordinary. Ideas are too brittle to accomplish this: "As an alternative, each saying can be heard as a call to go 'beyond' the saying—to activate the knowledge in what is said." (63) To activate we must envision, and to envision we must dwell.

Knowledge proceeds by challenging reference points. "As an arising, it is also a turning inward and a going beyond." (64) The structure of the chapters that follow proceed in this way. They are not linear in their presentation. Connections may operate at a more subtle level. Withhold judgment.

Chapter 1 67–76

The Availability of Knowledge K2

The taking form of experience does not exhaust the creative power of knowledge. It remains available even in the repetitive cycles of thought and emotional reactivity. It appears in attenuated form in the infinite range of conceptuality. Appearance owes its origin to knowledge, and without knowledge it would disappear. But consciousness imposes structures that conceal this creativity in favor of the claims of substance. Still, access remains available, most freely in inquiry. "When questions aim at what is meaningful instead of what is interesting, their aliveness offers knowledgeability." (69)

Today knowledge is stirring in new ways, and the future is less specified than ever before. We can assemble and package knowledge, but knowledge regarding our human destiny remains unavailable. To change this, to draw up new blueprints, we need a precise way of proceeding. We need a different way of questioning, one that engages head and heart; that illuminates the patterns of our experience. Once this dynamic is activated, even questions that seem abstract draw out knowledge and invite transformation. They are the starting point, the path, and also the goal.

Questions that aim at 'getting something' do little good; questions that jump directly into issues of meaning and purpose remain too abstract. Hence the importance of inquiry into "time, space, and knowledge as active agents in our lives. Such an inquiry engages the substance of our being without committing us to the structures into which we shape that substance." (72) We can discover the depth and immediacy of time and learn to open space, to activate knowledge. The decisive shift is to realize this potential is available. But we must also change our mode of inquiry. Space gives us nothing to bite into, to anticipate in advance; time's power unnerves us; and knowledge seems boring and also hard work. But these views are all based on the concerns of the self, a focus TSK inquiry challenges. TSK reclaims the place the 'I' has claimed. It conforms to no accepted disciplines of inquiry, without becoming too exalted to be of value.

Chapter 2 77–90
Revelation of Knowledge K2/K1/T2

Knowledge reads out the ways we think, the conclusions we arrive at, and the structures and ways of knowing we accept. It does so in response to its own dynamics and qualities. We could say it has a purpose, but this is misleading; instead, there is a rhythm of emergent qualities.

We rely on the knowledge of technology. But is this the highest value our age has to offer? Its patterns of thinking are already old, and it holds the threat of harm. What is more, its benefits are limited. A deeper understanding of our own nature might offer more. Technology aims only to accomplish ends specified in advance, leaving unchallenged the steady stream of desires, which can never bring fulfillment.

This seems a design for failure. Then could we function differently? To investigate, we must look at the workings of the mind, but to do this our only tool is mind itself. Then is there a way of inquiry not bound to mind? To challenge mind as knower, we must challenge self as owner. The alternative to aim for is that knowledge infuses all appearance, shaping what appears: that mind and world alike express a more fundamental knowingness. This prospect suggests another: We can invent a new mind. Here the technological model, which only serves the self, must be left behind. The power of our our own presence and potential can inspire a different inquiry.

We can discover a different knowledgeability in space and time: space within substance; time as the carrier (and nurturing flow) of experience; knowledge as embodied. "Our usual understanding of time, space, and knowledge as blank abstractions mirrors our estrangement from our own being." (87–88) When we let go of it, we have a ready answer to the charge that TSK inquiry confuses subjective responses with objective reality. When we let time come forward, experience is infused with time, and this means that when we investigate experience, we touch time. In the same way, we discover space in the way matter appears, and knowledge in the operation and formation of mind. TSK is thus more than metaphor; it is an invitation to recover the significance of time, space, and knowledge. What matters is to see how our current world arises through a tsk interplay.

Chapter 3 91–103

Dimensionalities of Knowledge K2/K1/T2

New developments that take place at the very center of the presupposed are the most likely carrier for the newness Body of Knowledge. We tend to turn away from this possibility, but need not. Our limits themselves express knowledge.

Subjective knowing arises within a greater not-knowing. The same not-knowing arises in the act of perception, where we can never reach what we grasp at. We try to control perception through labels and interpretations, but this only removes us further from the immediate object, while encouraging sameness. We move within the circles of the possible, shaped by previous mindings. We assign truth to the stories we tell, rather than looking to the actual experience. "The rhythms of knowing, the available imagery, and a kind of dictionary of possibilities conspire to shape the range of the knowable." (96) "Being here now" will not counter these trends or restore vitality to perception. It seems clear: There is no way out.

"Time, space, and knowledge [constitute] our own neighborhood. Our concern is with our humanhood, and our discovery is that spacehood, timehood, and knowledgehood may lead directly toward liberationhood." (98–99) We can proceed by bringing time, space, and knowledge into the foreground, enjoying their boundless variability, looking at the mechanisms at work, moving freely back and forth. "The more we conduct knowledge toward time and space, the more patterns and rhythms that give rise to alternative understandings emerge and interact." (100) Logic may tell us there is no way out, but logic is not privileged: thinking, speculation, and conjecture can strengthen inquiry. The process depends on seeing the impact of the lineage of the past. "We see the magnitude of the mechanism in play and sense directly its world-composing charisma." (102)

Chapter 4 104–118
Dynamic of Knowledge K2/K1

Conventional knowledge is impenetrable. "Caught in its fabricated world, we can only dance around [its] edges . . . anxiously weaving rumors into simulations of knowledgeability." (104) But within the Body of Knowledge, we know differently. Inquiry finds the small gaps and transitions.

Structures of conventional knowledge close down openness. Subject projects out to object, and mind is caught in its own images. But the medium of fabrication is only names. Interpretation reads interpretation; mind reads mind. The self emerges as owner, and specific ways of knowing set up limits. Rules beget patterns beget the structure of a system. Emotional responses are guaranteed. "The tightness of this structure makes a cage for knowledge." (110)

Our presupposed separation from the Body of Knowledge is an output of the system. It gives way when we touch the texture of space, experience the dynamic of time, and see in the light of knowledge. Inquiry arises naturally; we live more holistically and creatively. "In all ages, knowledge has conducted its own inquiry. The special value of the TSK vision is that it makes the subject under analysis so directly available." (115)

Chapter 5 119–25

Unexpected Knowledge K2

Knowledge is available within immediate experience: directly in the senses. This does not mean abandoning reason, but we need not stop with reason. We can learn to use the reasonable as a starting point for improvisation.

Our present, free of the way we shape it in light of the past, is indeterminate. There is always the possibility in the next moment of a different telling. Absolutes become obsolete. Such fallibility is hard to maintain, but we can remind ourselves of five limits on generalization. Speculation and wonder offer a different kind of precision.

Thinking is by no means the enemy. Thinking thinking 'right now', we discover that thinking insists on 'is'; it chooses between alternatives. These are limits that fall away in time and space and knowledge. "We start with the fundamentals of our body and mind as we experience them." (125)

Chapter 6 126–35

Actuality of Knowledge K2

When we are precise, knowledge can shine through all conventional structures, without regard to the specifics of the content we are pursuing. "It emerges as the beauty of truth, akin to the sharp and sudden presence of art—knowledge as the vision of knowledge." (126) If we are sensitive, "we may recognize that what had previously seemed solid and certain is born in the instant of inquiry." (127) Every form and design and category offers the illumination of an ever-present newness. "Entirely apart from the conclusions we reach, we activate knowledge in the activity of knowing." (128) And this holds true even if our conclusion is that we have no way to activate knowledge.

Mind seems more fundamental than its content: an open space ready to be taken up. But it is not passive either: it conducts, produces, and directs; it gives experience. It may be that it constitutes a vast unknown, with ordinary mental events only ripples on its surface.

Caught up in the content of our stories, we do not see that our experience originates in a time-space-knowledge telling. Connecting to this telling, participating in it, lets us stop insisting on order and certainty, limits and distinctions. "Whatever we say, our saying allows for the opposite." (133) All we need do is let go of the claims of existence, and we enter silent clarity, dancing and engaging.

Chapter 7 136–146
Opening Knowledge K3/S3/K1/K2

We do not have to rely on stories and sense projections, for knowledge participates in each manifestation, generating light. As mind inquires into this light, it discovers its own lightness in the readiness for insight, free from the gravity [grabity] of thinking. Where conventional space is devoid of knowledge, in the heart of space, the light of knowledge illuminates all.

When mind stays on its own surface, it can only go round in circles. One response is to touch the depths of mind directly, as in meditation. But then the link to knowledge remains submerged. If we stay on the surface, we do psychology or science; if we go within, we do philosophy or religion. But we are looking for something wholly different. The aim is to activate the vastness of mind *within* the operations of ordinary mind. When we stop thinking of mind as an object, we can see its fabrication in operation through a subtle introspection. One approach is to look at the self as a particular patterning in space and time, careful not to fall into the usual modes of identification.

The great conducting and self-exhibition of time, space, and knowledge give access to knowledge within each structure. Free from definition, "train yourself to raise the best knowledge you can." (145)

Chapter 8 147–56

Unknown Center of Knowledge K2/S3/T2

Shifting mind requires translation from the fixed to the open; requires entering an 'open trance' that crosses over to transformation, that transmutes and transmits. Trance is 'trans': an active movement into transparency. It is the entry point for knowledge, the point of transition that offers clarity and innovation. And it is transitional.

We sometimes touch vastness mind spontaneously, but such moments do not endure. There is, however, a way to rest in the feeling of openness, whose 'object' is mind itself and in which perception has no center, no specific source, and no precise object. Some ways to do this: Look at the space between thoughts; let appearance arise without location and history; resist the pull of the predetermined; disengage hope, fear, and memory. It is like developing an art of the mind, an aesthetic of awareness. Questioning is key.

The body of space does not make itself available to be measured, but space remains 'free-occupied'. It is there in sense perception as the echo; it turns existence into metaphor. The capacity for sensing that arises in time cannot be traced back to a first moment, nor can we establish with certainty that time has a forward directionality, or that this is the best way to link knowledge and time.

Chapter 9 157–64
Parenthood of Knowledge K2

Knowledge is usually thought to arise through an act of knowing, but we could look instead to the pre-existing 'body of knowledge' from which it emerges. Then the act of knowing consists in part in recognizing that we formerly did not know. Knowing and not-knowing are made of the same stuff. "In the search for knowledge, ignorance is not the enemy." (158)

We distrust and exclude not-knowing, but we could instead consider it our ally, seeing in it the creative source of the known. In that case, knowledge is not faced with the impossible task of establishing its own credentials, apart from not-knowing. Insisting on a sharp distinction is the real not-knowing. "Only on the surface does discrimination claim to be deep-rooted." (161)

In the play of experience, we discover the dance of knowledge. This can be our path.

Conclusion 165–77 K3

Knowledge has unfolded in history in countless ways, but our problems continue. The silent place at the center of our being, where knowledge is available, eludes us. We have learned to accept limits. But now we can investigate limits and past dictation, discovering "that there is really no turning away from the Body of Knowledge." (167) Knowledge is not just a tool or an instruction manual; it does not depend on routinization or meeting expectations. Rather, it is a part of being, and can come alive in our hearts and spirit. We can know as we breathe. "The Body of Knowledge can come alive and inhabit our world." (169)

Inquiry feeds knowledge into knowledge: It operates in all commitment. "It is the essence of serenity, the heart of faith, the spirit of science, and the guts of practical achievement." (170) We are always at the frontier, the borderland, ready to explore. We can start by no longer insisting on the importance of our own concerns: Then we can travel far more lightly. The point we inhabit can become the whole of a universal unique, available within our present livelihood. When we stand in the freedom of no fear, we can engage our situation fully and conduct inquiry without reservation. We are 'open-standers' and 'free-standers', no longer looking for fixed answers to fixed questions. We simply let knowledge speak.

Dynamics of Time and Space

Transcending Limits on Knowledge

Chapter One 3–8

A Challenge to Substance S1

If we expanded a molecule far enough, eventually there would nothing left: an expansion of appearance into 'nothing'. But the distinction is one of perspective, not sequence. Emptiness and form co-exist. Then is the world of substance only a condensed form of the 'nothing' we find in expansion? This seems wrong. Perhaps the answer is that there is an ultimate particle that cannot be expanded. But why should this be? In principle, whatever occupies space can be expanded.

We generally hold that whatever appears has a defining boundary. In expansion this border becomes porous, eventually disappearing. In the end, inside and outside can no longer be distinguished. Then what of the space within which the expanding took place? Does it expand also? Does it eventually also become nothing? This makes no sense, since space is already nothing. But that nothing is tied to the something that occupies it, so perhaps space would disappear after all. If this space transformation is the negation of a negation, do we then arrive at something positive; e.g., a more fundamental kind of nothing? Asking this has a great virtue: It makes space available for inquiry.

Chapter Two 9–18
Tracing the Origins of Reality S1

If matter became phantom-like, would the space that contains it change? How can we tell, when we cannot make space an object of inquiry? Through a different method, like opening a box by taking the box apart. In that case, matter disappears as well. "With no container, there is also no containing." (10)

To open space, we must make room for it. The best way seems to be to inquire into substance. Here, despite our experience, nothing seems fixed and definite (given the evidence of science). Experientially, we can mount the same challenge. Everything is in flux. This is hard to observe, because our knowledge points instead at fixed entities and identities. We might try to rely on trends, but these rely on taking up this or that particular perspective or concern. Consider, for example, movement while driving in a car.

Then perhaps the reality of the whole is more accessible. Somehow the whole 'seeds itself' into each particular, so that they can all be linked. Then what is its realness? Perhaps it traces to being a *meaningful* whole, derived from the interactions of co-referring manifestations. This leads us to the field and the field communiqué. Individual appearances manifest in terms of the communiqué, sharing and enacting a special intimacy. "The whole sustains the identity of each particular part, while each part in turn confirms the manifested whole... The 'source' of substance is the interaction of what does not have substance." (17–18) This suggests that subject and object are equally established or non-established; more broadly, that knowledge of what appears is given together with what appears (for the basis is communication).

Chapter 3 19–28
Returning to the Nameless S2/K2

The field communiqué transmits itself forward through names and concepts, giving qualities that are referred to entities. A world is set in place, and in this world we are identified as the one who knows. We know in a specific way, consistent with the logic of naming and the identity we have been assigned. Nor do words and images have a special status; they too have a role as carriers of the communiqué, bound to its mechanics and dynamics. Yet words can also evoke what is new, perhaps because they carry the power of the communiqué. Is it possible to evoke this power more directly? We could try the turn toward direct experience, but that is a difficult path indeed.

Perhaps we can turn to a different level, where subject and object, knower and known interact to produce the real. This ongoing interaction "pronounces the interior of the surface: the substance of what is and the 'isness' of the one who experiences." (22–23). Yet the interior pronounced in this way remains a surface phenomenon: an echo of the communiqué. To miss this point is simply to accept the truth of the story being communicated. And this is just what we do. We confirm our being, together with the reality of the whole, each a reflection of the other. We pronounce something as real. But this is to miss the point: the aliveness of the communiqué. "We have framed a reality that has no heart." (24) The result is that knowledge is impoverished. History is in large measure the account of the attempts to overcome that limitation.

Seeing the field communiqué for what it is suggests that limits can lose their finality. But for this to be so, substance must lose its substantiality. Then we can play with the reality of what appears, invoking a remarkable magic, "the other side of birth and the beneath of zero." (27) We can touch the nameless and the formless within appearance that make naming and giving form possible. We see that appearance does no more than trace the circle of zero, and return to zero.

Chapter 4 29–34

Inviting Space Knowledge S2

No-body space offers universal knowledgeability. But we have made space into nothing at all; have made it disappear.

Making it unknowable, we have made it disappear a second time. Now we want to restore space to appearance. But it cannot appear as object. Instead, we begin to find it as active within appearance, as the allowing of appearance. When we try to investigate this allowing, however, together with the mystery it reveals at the heart of all appearance, its significance escapes us; has disappeared along with space itself.

To evoke space appearance, consider appearance in space as space in operation. Merged with the echoes of the field communiqué, such space operation suggests that each appearance—in never taking form—fills the whole of space. Appearance remains indeterminate, and the impossible is preserved. Forms exhibit but do not take up. 'Is' and 'is not' are newly fluid.

Chapter 5 35–41
Space without Occupation S2/K2

The possible has taken form: that is so for us. But we can appreciate other possibilities, and this appreciation awakens potential. Properties can appear without possession or ownership, but also without being subjective; in fact, they must so appear. Space allows appearance, but it does not allow the body of what appears. Only in occupied space, the space of substance, is this allowing lost. In contrast, no-body space has never been occupied. The inwardness of no-body space is also the Body of Space, appearing through appearance.

Imagine that space projects inward to the mind the images of what appears. There must be a corresponding projection outward of knowing capacity, a projection that can in principle remain inwardly available. Call this 'inward outward-knowing' "eknosis." Here space, appearance, and mind all come together. The possibility arises of space knowledge, beyond the subject/object knowledge that can only move toward limits. Eknosis discovers within limits the aura of knowledgeability, the Body of Knowledge. "Knowing inwardly and outwardly in a single movement, eknosis moves without departing from its space origin." (41)

Chapter 6 42–50

Space Dimensionality S2/K2

If we could flatten space, it would mean flattening every point in space. Every point, already zero, would become zeroless. Would these realm be less real? Would what is left necessarily be 'nothing at all'? In any event, there would be a difference. Free from dimensionality, space would be free of the claims of substance [the "pronouncing of the interior of the surface."] Appearance itself would become zeroless, without occupancy or possession. The edge would open into fullness, and exclusion would vanish. "Every 'where' is any 'where', and each edge opens into the whole. Even the edge of the edge opens." (44) Nothing is established, and all is knowledgeable, in the special sense that it is wholly accessible to knowledge, supporting universal knowingness. In zeroless space, substance dissolves; that is, it appears as the appearing of substance. In the completely open, all is allowed. "What appears to be so becomes questionable in fruitful ways." (46) In opening to this opening, we can reverse space disappearance.

We do not inhabit zeroless space because the 'I' has taken up a position at the center, affirming and establishing. Yet this dynamic refers to nothing beyond itself. It can easily be reversed. If the zero remains true to the zeroless, there is no impulse to reach out; we could even say that dimensionalization has never happened. Appearance can be sustained as appearance, but because there is no dimensionalized substantiality, universal knowledgeability remains available. We can reverse the split into knower and known, into the 'zero' of subjective consciousness and the 'one' of objective identity.

Chapter 7 51–58

Thought Forms in Space S2/S3

The mind has structured a world in which space has disappeared. Thinking is its vehicle. But why are we always thinking? Thoughts seem to be the carriers for affirming fundamental interpretive structures: They sustain the world we inhabit by labeling it in accord with the certainties of identity, assuring the impossibility of gaps. They communicate the substantiality of their content, a kind of gravitational pull. From where do thoughts take this power? From the communiqué. Entering the world of the thought, we can no longer challenge its content. "Patterns of engaged images yield familiar projections, which are pronounced into daily life." (54) This familiarity confirms the known, but actually there has been no confirmation. Still, the resulting mystery appears insoluble. We cannot question thought itself. "In presenting their content, thoughts refer their own identity to this content. In the act of establishing, they withdraw." (54) Thought thus parallels space, presenting itself as a blank no-thing that enables 'something'. We cannot go to a hypothetical no-thought realm, but we do have the choice of noticing thoughts as they pop up—as sheer, sourceless appearance. Here thought, like space, is dimensionless and unestablished. Can we stay in that freedom? "Could we unknow the knowing of the known to know anew?" (57) Could we activate appearance as symbolic, making thoughts the agents of zeroless space?

Chapter 8 59–70
Interacting Magic S2/S3/K3

The dynamic that pulls thought toward substance manifests in stories and narratives that interact and dovetail. In each story, we assign new identities (even if we call such identities labels applied to an underlying identity). We affirm substance and close off inquiry into story-telling itself; in fact, the story of substance compels that result, turning stories into the zeroless and invisible, while translating space into dimensionality. Through the appropriation of knowledgeability, the aliveness of the story is translated into the solidity of what is told. As a result, the zeroless becomes unthinkable, space disappears, and mind withdraws. Knowledge vanishes into the domain of subjectivity, conventional space and time emerge, and not-knowing prevails.

Identified entities must be able to interact. This interactability must be viewed in conventional terms as a subjective phenomenon, a matter of interpretation. But it goes beyond this; e.g., the interplay of the senses, of subject and object. In the end, identities give way to inseparability, to multiple dimensionality—a gift offered by the zeroless, which assures non-exclusion.

Inquiry restores the energy of the zeroless. Through analysis, speculation, and imagination it frees thoughts from substance, and allows for pointless exploration. Knowledge itself is differently dimensioned. In zeroless space, we discover a 'finity' of dimensions, and leave the possible open and variable. Appearance remains space, and content arises as zeroless. Since the present has no parentage, conventional knowledge has no object. Yet knowledge becomes available everywhere. "Structures proclaim substance and thoughts proclaim content, but zerolessness reveals that structure and substance, content and story, are on the same level." (70) We are invariably 'within'.

Chapter 9 73–78

Time's Present Body T1

Experience unfolds through the three times. Yet past and future are unknowns. The fullness of time's temporality can only be a construct, for nothing in experience guarantees linear sequencing. We deal only with the present, but even the present is problematic. Is it only a moment of time? How does it link to the next moment? Is it only a shifting position, marking the interaction of past and future: a flowing state of 'being present'?

We seem to operate with both these views, as well as a third, in which the present is the construct of our immediate concern. But how does this square with linear time and the temporal dynamic? Do we know time at all, when we know only its passing away?

Chapter 10 79–90

Record of Time Past T1

The future is unlimited, but the past, as history, is inherently limited in scope, structured into various lineages. Objective time may seem unlimited, but that is only because of how we understand objects. And even objectively, simultaneous events have different, and hence limited, pasts. Past as totality is a construct.

If we want to isolate a point to explore its nature, we cannot take the present, since it is always shifting. Instead, we can take the point at the beginning, which can stand by itself. The present cannot stand by itself in this same way; cannot be identified. In the moment that we specify the present, we turn it into the past. All we ever find are past moments. If time is linear, all of time is past time. Even future moments will be the playing out of the past: they have already 'happeneded'.

We found our lives on pastness, enacting an identity determined in advance. We tell past-centered stories to confirm assigned identities and affirm a world of no freedom. The past accumulates layer by layer, and patterns increase in density. As we move toward increasing dimensionalization, emotionality builds up. We may wonder if we are truly alive. Are we ready to change this frustrating dynamic—to give up identity and ownership?

Chapter 11 91–101

The Future Infinitive T2

Time's momentum is a thrusting, driving presence, to which we all occasionally gain access. We can invite it by focusing on the future. In the future, the possible persists; the linear structure of time is unfounded. We miss this when we treat the future on the basis of past 'happenedings', initiating directions for future developments. We miss time's "transitional indeterminacy;" the 'future infinitive' of time. The future will never take form, and so is the source of freedom, best reflected in an active not-knowing. The future cannot be dimensioned; cannot be affected by the past. "'What is' appears in all its particularity, yet there is a 'seeing through', as though the well-founded conventional structures of past, present, and future had become transparent." (94–95) The future gives the present aliveness; it makes an ongoing offering of being. Poised at the edge of the present, grateful for possibilities, ready to accept the future's offer, we awaken to time's dynamic.

The future sustains the present through its never completed coming. It keeps going on, and is ever available. The present can be seen as the configuring of past and future (the meeting point of the symbol for infinity). Inhabiting the present, we inhabit the whole of time, and time inhabits us. "A forward allowing transmits the future to the present, allowing experience to remain 'in transit'." (97) We discover this 'transitional transmission' in our own aliveness, our 'going on'. "Experience is not the special province of the present; rather, it is what is transmitted as time 'moves on' . . . Whatever is is in transit; whatever appears can be transcended." (98)

Time brings the universe alive, freeing it from 'from' and 'to'. The future is the natural focal point for this dimensionless dimension, which is available in all the provinces of time. "To discover it, we can go directly to the point of arising, the point where the future comes into being." (99) Or we can look for other times of aliveness. We find that time sparkles and glows; concentration comes easily. Moving from *within* the future, we do not conduct 'from' or 'to'.

Chapter 12 102–114

The Present Timing of Knowledge T2

Choosing the past, we freeze time's dynamic. In part this freezing traces to our not knowing the dynamic, not being able to account for how time moves. We try to impose identities on the flow; we construct coherent narratives; we call on a witness. The witness lets us counter time's flow, singling out the present moment as substantial at the cost of restricting knowledge to that point, to the reach and range of the present. In the end, we affirm not knowing. Yet the claims of the witness in its limited relation to time can be challenged. The sequence of nows is simply conjecture. The problems go deeper: There is no time for witnessing and no witness for the witness. As momentary, the witness cannot affirm its own identity. And what of bearing witness to memory? More fundamentally, to set up identity and bear witness runs counter to time's fundamental dynamic. The claims of the witness are surely not false, but they presuppose too much: presuppose ownership and identity and dismiss the freedom of the zeroless.

The witness implies that consciousness as we know it requires substance and identity to know at all. But is this so? In a zig-zag pattern of time, we may be able to find other progressions 'embedded' within the old. Or we could experiment with reversing the linear flow. Such ways of knowing and interacting with time do not depend on the 'from' and 'to'. We do not need to 'go' anywhere. We can turn inward to discover the dynamic of time that allows change and transition. This is not a withdrawing into subjective experience (as the witness would have it), for 'subjective' is a construct of past-centered ways of knowing. Looking in this way, the substantiality of the witness gives way. Its claims to knowledge do manifest knowledge, but need not be accepted as true.

Chapter 13 115–30
Conducting Temporal Sameness T1/T2

Time cannot be known by an act of knowing or telling in time. But consider a different telling, one that evokes a no-story knowing: The Conductor's Story. Time has seeded the past in our minds; now we react, conducting order, shaping new rhythms in accord with accustomed patterns. "From one single thought, almost from one single word, we could conduct the whole." (117) We can only conduct theme and variations, based on the conducted time of objects and of self. One moment lines up after the next, like beads on a necklace.

Everything proceeds with great precision. The potential of multidimensional appearance is converted into a juggernaut. Everything reinforces its momentum, feeding the gravity of thinking and assuring a string of sameness.

Conducting in this way guarantees limits, like a police officer directing traffic. We seldom even notice these limits (linear time, self-centering, lack of focus, etc.), or see how not-knowing becomes central. We refine and rework old themes, conducting repetition: an echo world where voices repeat endlessly, heading toward limits, pain, numbness. We cannot distinguish between our tellings and what actually appears.

In conducting stories, we also conduct time—in effect, we conduct conducting by conducting the time that supports conducting. We conduct time as linear, as supporting subjective and objective, as marked by transitionality and as measured out. We interpret in ways meant to reconcile conditionality and conditioning on the one hand with identity and sameness on the other. We conduct certain ways of knowing, though other ways of knowing remain available. We set in place a knowing in which one appearance depends on others. In the end, we are forced either to eternalism (an atemporal first cause) or nihilism (only mutual referring). Both share the same understanding of time: "As the 'on' of what arises, time 'goes on' in ways we can 'depend on'. (129)

All this is built up by mind out of points and particles of interaction ('in' and 'on' and 'by', etc.). Yet mind is built up in the same way. It is like painting the canvas on which the painting will be painted. The meaning-giving mind depends on the meanings it constructs. This could be understood as the ongoing openness of this future, free of limits.

Chapter 14 131–44

Opening Unknown Time T3

The order of the whole encompasses the time that transmits the whole: the tempo of the order, playing out in different ways. Taking the microcosm and macrocosm as examples, the tempo is like the base for two cones. But what of the region beyond the cones? Here any temporality must go unknown, for it is not based on the tempo.

Within the prevailing temporal orders—in the projecting forward of the tempo and in other ways as well—unknown time is available to be known. We access it if we do not proclaim an exclusive temporal order. We can speculate as to what this unknown might be like; we can get clear on what it would not be like (no borders or projections, no established categories). We can affirm its necessity, since the temporal order itself somehow 'originates'. But it remains 'hiddened': out of view, before origination, while still wholly participating.

We can posit a deeper structure of time, in perfect intimacy with our being, interacting with awareness. Before the separation of time and knowledge, a new temporal knowing presents itself: both more and less than possible. To arrive at it, we must trace in a way that does not depend on the order within which we trace; must trace the arising of the order. Then we may be able to arrive at an unknown knowing, like unknown time, that founds conventional knowledge. There are hints that the 'prior' of this unknown time is available, if we can shift the 'heavy-handed' directionality of the temporal momentum. Then we can find the unknown prior within the ordinary. But we must first inhabit fully the ordinary. We can choose and cherish knowledge in all appearance. "The depth of our live *becomes* the inward intensity of time. . . We can awaken . . . the universal unique." (144)

Chapter 15 145–55
Dancing a Challenge to the Whole T3

"Can we conduct into awareness the trend toward transformation that flickers at the edge of our truths?" (145) Asking the question generates new questions about how time operates, about the fluctuating dynamic of ongoing 'presencing'. "If we chose not to place identity at the center of our conducting, we could allow the dance of time to float into our imaginations." (146) Imagine that instead of identity, there is the dance of time. Things are and are not, in a celebration of finity. "When the story is its telling, appearing is its appearing." (147) We can conduct as a wire conducts. We draw closer to spontaneous knowledgeability.

Consider the unknown time that allows conducting as nuclear time, going nowhere: the not-not-not. The nuclear point becomes the universal unique. Conducting dance rather than narrative, without identification, we can improvise. Improvising instead of projecting, we can return to full seeing, replacing dimensionality with a way of seeing based on light.

Projection can proceed differently, practicing instant reflection. Seen in that light, even our present projection is infinitely skilled. The imperfections (pain, ignorance, obstacles) are the opening into what is intrinsically flawless. Improvising on the basis of imperfection, knowledge presents the universal unique: an inward improvised conducting that arrives at the subtle 'inward' of time. "We touch the lightness of being; we fuse with light." (154) Here words fall short. In this inaccuracy of saying, logos takes on a different power: potentially the source of life of the knowledge of being.

Chapter 16 156–69

A Bridge to the Body of Time T3/K2/T2

A gateway to nuclear time is impermanence, viewed in its subtlety as insubstantiality. Then we see the mystery of time's presentation pervading experience; we enjoy an inward knowledge which rejects any attempt at mystification. "In this ongoing going on, mystery itself is rededicated as the truth of experience." (159) We return to eknosis, turning the mechanisms of temporal presentation inward, de-dimensionalizing time. "The temporal mechanisms that give order to appearance appear *'alongside' the things or events they purport to order*." (160)

Eknosis reveals the forward-projecting recordings of the conventional order as the prerecorded contents of other recordings, with the mechanism that arranges for pointing back and forth among recordings appearing 'alongside'. Linear time is just another mechanism. An unlimited 'rhythm expression' offers the Body of Time, pointless, yet presenting all points. When we inhabit the inwardness of time, a different dynamic is available. An unrestrained play of emerging arisings.

We can imagine a nameless arising, a propagation whose record founds the genealogy of what appears. Recorded time goes forward, giving a way that works. "Each topic . . . is the segregated subject of a seamless instigation." (167) Yet the inner time of recording is not itself recorded. "In the persistence of the possible, the boundless creative presents." (168) To cross over into possibility, we need a new experiencer: an experiencer transformed into light. "Silent light penetrates the surface of time." (169)

Chapter 17 173–83

Knowledge in Time and Space K1/K2

The bystander model, grounded in not-knowing, determines our knowing, which proceeds by forcing experience into predetermined molds. We become bias-standers. insisting on the models we have fabricated and insisting more fundamentally still on dichotomy. Deeper knowledge and visionary knowledge fall by the wayside. Relying on 'from-to' derivations, we have no choice.

We can let go of the job we have taken on. Time and space can speak for themselves. "We do not have to proclaim our portion, because we are already inseparably related to the whole." (178) We can merge with a transcending knowledge. We can learn to listen to knowledge as it has manifested in history in all the domains of human endeavor. Knowledge is knowable, not as something to be known by somebody, but as intrinsically knowledgeable. Setting aside all judging, feeding knowledge into knowledge, we arrive at wisdom. Turning inward toward time and space, we connect to ourselves and to knowledge; giver of wings and worker of wonders.

Chapter 18 184–98

The Priority of Knowledge K3

Conventional knowledge carries the Body of Knowledge, available in the subtle mechanisms of knowing and also in sudden shifts—internal revolutions. To gain access, we must challenge the bystander self and the patterns it sets up: the patterns of kangaroo mind. We can start with the model of knowledge as arising in time. What if knowledge is 'prior'? Then the field of knowing as such can open, without depending on positions. The prior can become primary. We do not need to trace a source (which at once places a limit on knowledge); we can question the 'source from' perspective, going beyond the zero-point of origination to a knowledge that is zeroless. But the going beyond must also stay within. It may seem that if we stay within, we remain subject to limits, but if there is no source from, and nothing has ever arisen, there can also be no restrictions. "Whatever we do, we are not caught." (193) Knowledge conducts a zeroless that cannot be confined. "In place of birth there is apparition; in place of creation, there is the display of magic. (194)

Knowledge is routinely reduced, but we can search out the exceptions. "Knowledge of the whole—the inward knowledge of a rainbow eknosis—is available in the obsolescence and inadequacy of our current not-knowing." (195) The place to look is to the 'not' of this not-knowing. Then we discover that knowledge has to do with love, intimacy, and joy.

Chapter 19 199–208

Light of Knowledge K3

Speculating, we can say that knowledge is linked to light. We could imagine an unformed light radiating through space prior to dimensionality: a fundamental light communication.

From a light mechanism, there is a transition to the physical, with hydrogen marking the point of transition. A 'going toward' expresses time and activates space receptivity: light as the instrument for creation. In the mental realm the light mechanism presents reality as reflection. In both ways, light serves as the source for appearance, but not in any conventional, 'from-to' way. In inquiry, we can trace this light transmission, starting with physical light and moving on to the 'lightness' of sensing.

As the light momentum accelerates, it approaches the zeroless. We discover countless openings in all appearance, a "secret biology of luminosity." We discover what it means to embody light. "Meanings become precise and insight penetrating. . . . In the beauty of this light projection, there is a power that approaches magic." (206–7) We may find ourselves speaking in the language of the old religious traditions, of love or blessings or enlightened consciousness.

Chapter 20 209–14

Darkness into Light K2/K3

For bystander knowledge, light reveals objects to a knower, and this means the knower looks away from the light toward what is revealed, living a 'no-light' livelihood. We bathe experience in a kind of 'darklight'. The alternative is to invite light, starting by fully inhabiting the situation in which we find ourselves by sinking 'beneath' the contents of mind. We enter each position, fearlessly conducting forward. Pursuing this course, we eventually *become* light. "At this subtle mystery level, light itself is only the appearance of light." (214) 'From' and 'to' fuse.

Chapter 21 215–23

Enacting Pure Knowledge K3

Light awareness offers a way of transforming whatever is negative. We can be confident in the 'lightful' purity of knowledge, knowing that each appearance leads to the universal unique. This does not mean turning toward the light and rejecting the dark: Our turning is a turning inward. "Time's rhythm as it plays itself out in our stories directs us toward the emotions and the sense of self, and there is no need to resist." (218) We are free to enact light transformation. And when knowledge has arrived at self-luminosity, it needs no more points of reference. We are free to pursue special states or attainments if we wish, but then we lose contact with the Body of Knowledge. "At the heart of knowledge, 'transcending' is transcended." (222) There is nothing to be done.

Chapter 22 224–27

Radiant Knowledge K3/T2

Knowledge requires no path; it transcends transcending. Yet still we move toward a transcendental: This too is part of the play. Knowledge founds the realness embodiment, predicting and predicating what is and is not. We can aim to make the knowledgeable self-projecting, so that it is wisdom that manifests. Or we may fall into constriction, so that the point is lost. The radiance of knowing dims. But limits call out for transcendence, and we are the architects of our own response.

Chapter 23 228–34

Treasures of Knowledge K3/T3

Knowledge accompanies all perception, expressing in a certain quality of what is perceived its own availability. One way to arrive at this embodiment is through opening time: first by cutting moments, and then by departing from the one dimensional arrow of time, toward the presence of the present. This opening can continue, toward an infinity of knowledge.

In going inward in time, we arrive at the universal unique. "The center becomes centerless: focused without focusing 'on' . . . A vastness opens within the particular." (230–31) The particular enters the unity of knowledge, which evolves toward time-space intimacy. Light leads to fullness, to an 'inmerging emerging'. Every point is vast.

Chapter 24 235–42

Commitment to the Whole K1/K3

Fabricated knowledge is not knowledge in the deepest sense at all. Sensing this, we become seekers of knowledge. But now we depart from knowledge, distracted into collecting knowledge as we are distracted from sound by the meaning it carries. "We have only been collecting the traces left behind by knowledge in its passing." (237) In the world inhabited in this way, the Body of Knowledge has yet to be born. Now is the time for transformation. Is newness truly unavailable? "Knowledge the Old Father invites us to grow to full maturity." (240) We can act on what we know has value, abandoning our models and letting being itself become the model. We can release the time of conditioning that shapes our world.

Chapter 25 245–57

Freedom for Knowledge K2

Instead of accepting the substance of what appears, we can develop visions of knowledge, changing knowledge from servant to provider.

> *Exercise 1*: Look within each arising moment for the quality of awareness it carries; then look within awareness for knowingness and the sense of presence. The aim is not to find new knowledge, but to conduct openness.
>
> *Exercise 2*: Rest in the activity of thinking. See without the eyes; understand without the words. In this way, you can let go of the anxiety that thinking generates, together with commitment to the narrative.
>
> *Exercise 3*: Look for a characteristic numbness or heaviness or gravity of thoughts, and then begin to

shift them toward greater playfulness. You can cultivate a quality of inquiry even within dullness. Practice releasing the energy of thoughts into knowledge.

Chapter 26 258–66

Vastness of Experience K2/T2

Thoughts leave residues that shape the next arising. Echoes echo, becoming muffled but preserving the whole. The alternative is to let knowledge inspire knowledge, learning to expand questioning and return substance to light and space.

Exercise 4: Look at the arising of words, of thoughts. Look at sameness; at the one who knows; at mind without words; at the mechanism of conducting; at the pursuit of knowledge.

Exercise 5: Abide in thoughts, invoking the dynamic rhythm of time. Eventually you can cultivate this abiding within the flow of linear time. Consider abiding as a going 'inward'. Let images prepare the way for the dimensionless.

Exercise 6: Relax into 'no-conducting forward'. Expand each sensory capacity as its own world.

Chapter 27 267–76

Challenging Structures in Space S2

Knowing objects instead of space, we will never know for certain, for the object is unknown apart from the projections of the knower. The measure we apply shapes what we find, confining knowledge. Yet the measureless must be the source of what can be measured. To see this, consider that the distinction between 'is' and 'is not' may be incomplete; for instance, that we *could* and in a sense *do* inhabit multiple universes. Language conducts sameness, but that is deceptive: in each moment, countless variables come into play. "'Substance' proves to be an interpretive arising," (270) not opposed to space. Space proves available within the measured-out realm: a generative source. Reality becomes much more transient—just popping up or projected.

Exercise 7: Focus on an object and its surrounding space, and then expand. Shift perspectives, or try several perspectives at once. Expand and condense at various levels. Do the same with time.

Exercise 8: Imagine traveling in space with no direction. Just stay with the sense of space.

Chapter 28 277–84

Silent Appearing S3/K3

Instead of focusing on what has arisen, we can focus on the arising, training mind to gain access to the Body of Knowledge and awakening awareness.

Exercise 9: Look at space qualities such as distance or 'between'; let this more spacious aspect enter sense experience and thought.

Exercise 10: Look for an underlying continuum of stillness within experience: the womb of space. See how experience arises out of it, noticing how mind builds up experience from within stillness, without taking on the heaviness of the content that accumulates. In this way mind becomes more like open space, granting harmony.

Chapter 29 285–94

Coming Home to Time T1

Time is the hidden dimension, present within our stories as an active force. It is present in the flavor of each story and as a theme in our ongoing action, as well as in our inevitable grasping. We encounter it in alienated ways: in our habitual patterns and our explanations and justifications, in daydreams, emotions, and obsessions. Encountering our lost connection to time, we may build a precarious nest, anxious not to face our true situation.

> *Exercise 11*: Investigate the stories you tell. Look for second-level stories and for the transitions between stories; try to touch and release the energy bound up in stories.

> *Exercise 12*: Reflect on the patterns of your life, letting pain penetrate the power of the fabrications to feel time's passing. Cultivate caring, balancing it with analytic insight into the structures of time.

We seem cut off from the dynamic of time. Our livelihood lost, we back into the future, always being taken by surprise. To feel this loss evokes the clarity of knowledge and freedom of space.

Chapter 30 295–311

Healing Time T1/T2

Time's momentum, tempered by our sense of loss, results in a sense of time flickering. It can also lead to a more dramatic freezing of time. Spreading ourselves out across time, moving from point to point, we have the sense that what we do is pointless. Behind time, squeezed by time, we feel a characteristic pain, but because we identify with it, we cannot awaken to it. But within that pain, time's power is available.

> *Exercise 13*: Bring aliveness into experience, relaxing, and practice stretching time out, so that moments do not race by. Let the pain of temporality release, so that knowledge becomes more available. As a different temporal rhythm presents itself, we have greater access to knowledge. Time becomes, in a quite precise way, pointless. This too can be brought into the linear flow of time.

Exercise 14: Let time's dynamic manifest. When pain manifests, go to the 'painness', before labeling. "You may find that in the midst of painfulness, you can enact a time of purity, awareness, and light." (303)

Exercise 15: Try to use the energy of time's pain to penetrate the surface of time. This brings you closer to the heart of time. The truth of painness offers ways of healing. Balancing in the pain cannot be done by the self; that is the reason for focusing on painness, which offers pain without solidity. Awareness can arise as knower.

Exercise 16: Let pain into awareness, with the intention of healing but without identification. The shift is from content to communication of content: the open of no concepts and no doing: the silent communiqué of time. To expand this, look to the energy of thoughts rather than their content.

Chapter 31 310–20

Past Conducting T2

We do not have to overcome old patterns and obstacles or make great effort, for then the past becomes a burden. Instead, we can conduct the conflicts of the past inward. When the past is stripped of its weightiness, the future can present itself now, and we can embody what has come before. "Seeing becomes being, and we enter light everywhere." (312)

Exercise 17: Make the commitment to turn away from the negative; then look in your own life and in history at how the past has conducted knowledge forward. Note how limits arise and time is wasted. Ask how to go beyond and not duplicate.

Exercise 18: Review the past for mistakes, etc. and trace out the consequences. Involve yourself deeply. Look in terms of human history also, to see how knowledge unfolds and also is thwarted in time. Trace the patterns objectively, but also with great caring, letting objectivity and caring meet at the point of inwardness. Find points in the recent past where you rejected knowledge. Instead of explaining, let present and past feed each other. Look for a more encompassing knowledge available in the past and also the present. Let the past be a source of knowledge.

Exercise 19: Review the history of a present situation, looking for influences and forces. 'Rewind' the past and play it forward in a more positive way.

Chapter 32 321–39

Activating the Time of Knowledge T2/K3

Accounts of our time that view it as the Kaliyuga suggest that time can stand in different relationships with knowledge. The same is true in our own lives. The sense is that the characteristics displayed by objects *in* time are in some sense properties *of* time. Can time manufacture such properties? Can we influence this?

Living in the desert, we dig for water, for nectar, for treasure. But it never occurs to us that we need not live in the desert at all.

Exercise 20: Invite knowledge into the senses. We can ask new questions: "Is 'they are' not there, or is 'they' not?" (326)

Exercise 21: Invite into the presence of space and time a quality of light: centerless luminosity.

Exercise 22: Learn to be aware within darkness; discover the availability of light there. Look at the light in which a mental image appears. Look at the same quality of light in thoughts and feelings. Even in the darkest times, look for such light, which is never far away. Add a quality of joy and delight. Look at the places darkness most readily persists.

Exercise 23: Explore feelings of unknownness, expecting nothing. Time itself, the temporal flow, is a great unknown, but somehow it manifests, somehow it transmits knowledge. Allow the future to be present as completely indeterminate. Now look at and experience decay. What quality of time supports this? A kind of innocence may be available.

Exercise 24: Invite into each sense perception and thought the *wholeness* of perception. Appreciate the time-space interplay and your acting into it. In the face of these dark times, we may not have a plan, but we can take the first step.

Exercise 25: Imagine your life moving from light to lighter. Move into the presence of knowledge, as a master of knowledge. Note what holds you back. In the availability of appearance, the future is available, and unknown knowledge can manifest. Perhaps it is not as difficult as we imagine.

Sacred Dimensions of Time and Space

Introduction 3–14 K1/S1

We start from 'I am here'. This beginning gives us a zero-dimensional world, a zero-point in time and space and knowledge. But 'here' requires 'there', and now we have subject and object, self and world, as well as the line between them. But to locate points on a line requires axes. The axes have four end-points, each of which has four end-points. We arrive at 16: an indicator of completion. The zero point can also be opened up, along the axes, giving a cone, or 16 cones in all. But the whole structure remains reducible to zero; constitutes a reflection of zero. From this basis, the arising of forms and the activity of knowledge can be established and reduplicated, creating intricate structures, all based on radiating cones and zero-point spheres. This model, while not 'true', allows for a far-reaching inquiry not bound to the usual assumptions. "And because this inquiry is founded in 'I am here', it is one that any 'I' can make." (5)

If we focus on the subjective I, do we give up on objective knowledge? The link is closer than this question suggests. 'I am here' is the lens through which we look out on the world, and through which 'objective' knowledge becomes available. So we need to understand it. But at the same time, we do not wish to focus too closely on the subjective. This present approach does not fall into depending on the usual subjective structures. "Perhaps it is enough to say that we are investigating 'here' as well as 'there', that our aim is to clarify the 'from' from which our ordinary knowledge proceeds." (8) What we are looking at is like a play put on by time and space, with an unusual cast of characters. Watching, we begin to open the cone of subjective knowledge into zero, liberating energy and activating knowledge.

This space-geometry takes us to the background of our embodiment, the space-allowed 'there' that is the precondition for the 'here' of 'I am here'. The zero-point represents space-openness, and the unfolding from zero to 16 parallels the play of appearance. 'I am here' clarifies the intimacy of human being with time, space, and knowledge, and thus seems a fruitful way of telling the story. We start at the first level, but then a second-level openness begins to operate.

Space 1 17–24

Geometry of a Living Universe S2

The 'I am here' is our fundamental way of triangulating knowledge, time, and space; it founds the ratios of conventional knowledge that construct a known world. It seems self-evident, but it also seems worth challenging, given where it has led us. So instead of the usual interpretations, we can stay close to the fundamental structures. We see that existence seems related to location; that here and there provide points of reference and movement 'from' and 'to'. At the same time, 'here' is no fixed place; wherever I go, that is 'here'.

'I am here' suggests the structure of a baseline, but only on the basis of points whose position is determined by axes. Then we have a point of origin, a zero-point. We cannot say from where zero emerges, except to say it is the outcome of two-fold directionality; of two lines that intersect. In this we soon arrive at the fullness of 16, and the endpoints of the axes leads us there as well. Zero also points in sixteen directions.

"It is as though we had found sixteen different 'heres' within the 'here'. (24)

Space 2 25–27

Images of Wholeness S2

Identifying 'here' and 'there' without assigning them content is like giving a voice to space. We can trace space in the zero, considered as a two-dimensional circle (a circling back of the baseline that suggests the feedback that establishes the self, here, as the knower; compare the curve in the '6' of '16'). Next we can imagine lines converging on the zero in three dimensions, making the zero a sphere. This gives a more 'full' sixteen, as do the sixteen points of an imagined four-dimensional cube.

Space 3 28–31

Cone of Possibilities S2

We can imagine the zero radiating out as a cone, thus introducing time in the dynamic inherent in 'from' and 'to'. Cones and points alike can serve as zeros. A structure of vast complexity emerges, and the world takes form. But we can continue to look at structure rather than the specific forms that emerge. We can take the Giant Body as an example. "As the structure of multiple cones becomes more clear, the body, its aliveness, and its functions can all be seen as instances of space-time development: translucent points of being held together by the 'glue' of space." (30) Is this just wholly imaginary? Better to leave the status of the lines open. The very openness of what defines helps convey something about time, space, and knowledge: an openness that is ready for anything.

Space 4 32–37

Magical Body of Space S3

From 'here' and 'there' and 'from' and 'to' a space dynamic appears to allow for shape and form. Perhaps this dynamic or something like it functions in conventional space. This prospect or vision offers an advantage over conventional views, because it is open to the aliveness of appearance, unlike frozen conventional time. It links that aliveness to precision and specificity, and offers possible access.

Consider the zero-point as founding what appears. Clearly it cannot found substance, but we can think instead in terms of motion and location, which move and build without ever departing from zero. "I am here' builds and structures, relying on knowledge. This view seems more true to the openness of space. The zero-point can be seen as a symbol for space. This makes sense if zero can and does appear, for it would appear in a way that is closer to symbol than substance.

If zero is the point before it manifests, 16 is the foundation for the architecture of manifestation: the power of space to locate and allow. In this architecture, no point is wholly separate, for all points represent the 16 of the whole. "16 is the abstraction that becomes the fact." (37) It is a kind of code that reveals its own meaning in the convergence of the whole. "It is the point of transition and the transition of the point . . . It is the gathering point of time, space, and knowledge." (37)

Space 5 38–44
Manifestation of Form S2/K2

As a way of moving from the geometry of structure to the infinitely complex forms and strongly specific interactions of the being-point, there is ratio: the relationships that make up the whole. Ratio gives distinction, precision, and clear form, sameness and difference, mind, boundaries, and directionality. Ratio is living and dynamic, articulating change and transition. "When 16 reaches completion, ratio is what sets manifestation in motion, making transition possible." (41) Lower knowledge turns ratio to judgment, but that is not inherent in ratio, which is open and creative, manifesting a second-level knowledge that recognizes ratio as the perfect manifestation of space. Knowledge knows how to operate the ratio: It makes its point without being bound by it.

At the zero, time and space do not act. Change comes with the ratio, which engages the completion of 16. Every 16 is a ratio brought to fullness, a transition of the whole. Without the ratio, every point engages 16 dimensions of time, space, and knowledge, but with the ratio, 16 zero-points unfold together, manifesting limit, form, and meaning. At 15 the whole is available, and at 16 the 'right' ratio forms. A definitive whole coheres. But each step from 1 to 15 may have its own 16. We could imagine the unfolding as taking place in a fourth dimension.

Space 6 45–49

Space Performance S2

Space is the prior of matter's existence, available as matter's other. But what is the prior or other for space? Asking this question, some aspects of space emerge: space as partner with matter, space as linked to time. "This triple interaction of time-space-knowledge is also the juncture of our human being." (47) Here we act into the unfolding of what appears; here we move toward our destiny. But what is space? Space allows form, but remains unaffected. Space allows points, but cannot be a point of the whole. If each 16 draws on space-openness, are all points one?

Space 7 50–60

Structure of Zero S3

How can space allow? How can zero serve as symbol for its allowing? Zero is a line, but so is every number: space within, space without; edge without substance. We might think of zero as a remark made by space, making it both point and pointer, authorizing zero by decertifying all points. Mark becomes logos, and being unfolds. Zero is the place-holder, but also the founding zero-point. It neither is nor is not. It is sheer reflection, but so is all manifestation. Once structures are established, first-level knowledge cannot see this to be so, but it remains the case. Space says 'may' and time says 'can'.

First-level knowledge reads the 'can' as 'cannot', but the 'open' of space gives points as zero-points, location as possibility, and thus restores the 'can'. We build walls, but only by making use of zero. In depth we discover the missing connections. In the end we discover that we are bonded to space and time. There is always another direction, allowing time's momentum. "History and existence have space permission. Nothing holds them back." (59) Points appear and work distinctions, but they return to zero-creativity. The point of being carries knowledgeability.

Space 8 61–65

Space Qualities S3

Through space openness, mind relaxes. Everything connects: space to space and heart to heart. All edges are united by deep lines of space. Meaning and love awaken in the perfection of 16. In love with space, we find protection. Space teaches knowledge, the point of every point. "In the house of being embodied, all can become space." (64)

Time1 69–78

Life Time T1

We live in time, and time is everywhere dynamic. Yet we fail to notice this dynamic. Traveling through time, we do not touch its flow. Time develops into 'from' and 'to', before and after; it is timed out in sequences, from geology to biology to history to ideas. Themes repeat themselves, and events proliferate, filling up space.

We do not know how the sequencing operates, or how its transitions work. We do not explore its qualities or dimensions, its consistency and its changeability; we do not even think to investigate. Because we do not know time, we have not been able to develop incisive knowledge. But why do we accept the knowledge that guides us now? Is there a different path for knowledge?

Time 2 79–88

Ordinary Time T1

Time seems to move too quickly for us to take hold of it. Or is it that we are moving and time stands still? Time depends on space and place, on where we are in the world or in the universe. In outer space, what measure would we use? How do these measurements relate to the movement in experience, which goes on even in moments of perfect stillness? Our thoughts move, but how does this movement connect to the physical movements in the universe? Is measured time manifold, depending on our concerns?

We look at the past in terms of content: memories, etc. Can we look at 'pastness' more directly? What makes memory relate to the past: what feel of the image makes it so? How are memories oriented in relation to one another? How does the remembered moment differ from the present moment? What is the edge between past and present? Can we discover in such inquiry the dynamic of time, apart from its content?

What is the future, if not imagination and prediction? We imagine the future as an image that has a special 'not yet' feel. What gives it this forward projection? If we look by focusing on the edge of the present, where the future arises, there is both change and continuity, a holding and letting go. Can we touch the power of time in this way? Experience future and present at once?

Time 3 89–96

Space-Time Partnership T1/S1

Space accommodates time, while time brings vitality to space. Time opens the dimensions of space, giving depth—visual space is completely full and also flat. Only when I move do I walk *through* space. What if we moved differently in space: Would time shift? What if we could move through time as we move through space? It seems that time must be connected to space, as one moment of time is (mysteriously) connected to the next. Conventional knowledge confirms this. Perhaps space and time are not separate at all. Every point is a point in space *and* time. We say, "I am here now."

For time, the openness of zero means that everything is possible. Time's rhythms reflect space's repeating structures. We might say: the 16 of space joins with the 16 of time to produce the 16 of what is. But time and space synchronize completely; between them there is no 'between'.

Time 4 97–103

Endpoints of Time T1

To establish a point in space, we also need to find the baseline in time. We go from now to then: baseline is also timeline. We measure out time as we measure space on axes. But the endpoints that give the timeline are not arbitrary: one of them is 'now'. Is the other point the beginning of time? But how can we ever specify that point? If that point is the zero-point, it has been the same point through all of time, even as the 'now-point' steadily changes. Will there also be an end-point? Or is time constructed very differently? If the starting point is the zero point, will there be a negative timeline on the other side of zero?

We can imagine time's flow moving in two opposite directions: either the future arrives, or the future arises. If we switch back and forth, which gives a better sense of 'pastness'? 'futureness'? 'timeness'? Imagine running the flow of time backward: Would we come to a stopping point? Out of what would that point be manufactured? What could call a halt to time? If we imagine history as a cone, emanating out from a zero-point, why is there only one cone? What of other lines of time?

Time 5 104–11

Back to the Beginning T1

A starting point for time is troubling. Whatever we point to, there will be a prior. But how can there be a prior without time? Does the logic of 'to' and 'from' betray us here? Do our questions about time resolve into questions about the mind; about our projections? What is the ratio of our mental constructs to time itself? Could our not-knowing open another angle on 'to' and 'from'?

Existence and experience require time. In our life, time is limited. "The deadline cuts through the timeline like a knife." (108) But where does time get this power over us? We return to a different beginning, not billions of years ago, but now. 'Then' is not 'now'; it is the not-now whose relation to now determines the nature of now. "Every era tells its story to arrive at the 'was' of its 'is'." (111) And the self too is timed out across time. Yet it is now that we are and know. I *imagine* being there and then. Knowledge and time intersect at 'am'.

Time 6 112–17

Dimensions of Time T2

Suppose that any here and now could be a zero point, a starting point. The zero point is also 16, is also whole. And it must be whole to be a seed for what comes next. But what does it mean for time to unfold in sixteen directions? Time hides even past and future, presenting only partly a highly restricted one-dimensional structure. Could we discover a threefold dimensionality? Where would we look? Perhaps at the beginning, middle, and end of each moment, or the past, present, and future of past, present, and future. Would seeking such connections lead to a fourth dimension?

Think of a moment of time as a triangle, each angle marked by another triangle. We move from one apex to another, via an unknown mechanism. Are moments linked on a thread? Held together by a force field? Do they overlap? Grow out of one another? We cannot say, because we have no experience of the 'between'. Perhaps in that 'between' there are more fundamental points; perhaps we could dive down into those points. Or what about moving off at different angles?

Time 7 118–23

Depth of Time T2

What exists 'takes' time. But it does not take it up exclusively, any more than events in space take up space exclusively. There are layers of time and space, and at different layers, different things are happening. Linear time resists layering, but it may be that events in one spatial layer (atoms, molecules, etc.) operate in a very different time from our own. Could we enter different layers by changing the focal setting? We might develop more concentration, or might dive into time.

In a cone of time, both ends are zero, holding the line of time open between them. What if we could open the zero: would the cone unwind? What if we opened the angle on the cone by unfolding it, like an umbrella? What if we rotated it completely around? Would it disappear completely? Consider 16 lines of directionality and 16 cones emerging from 16 zeros; why do we inhabit only one cone and one direction? Imagine cones within cones, each with rings of 16, sharing eras of time. Would the operations of time necessarily be the same in different rings of the cone? How would the rings be connected, so that time could flow? As the cones unfold into the present, what becomes of the past? It is there in the structure of the cone, but does it *exist*? Does it simply vanish, as the past itself always seems to vanish?

Time 8 124–37

Power of Time T3

We experience time in gradual transitions, coordinated and connected. Old forms vanish as new ones emerge. But how does the power of time accomplish this? How does it transmit along cones of time? Transition points must also be zero points, so that a new pattern can emerge. Then do we start over in each moment, flashing off and on? Consider the cone as the ratio of length to face: Does manifestation depend on achieving a certain ratio? Or instead, suppose that it takes *no time* to move through the cone, to move from zero to zero. Could there be simultaneous development? Is there perhaps another time in which simultaneous and linear are not contradictory? Suppose each point requires 16 transition points to reach completion. The connections would have to be precise. If so, time's power could manifest there. But this would mean 16s everywhere, all the time.

Zero as zero has no properties, and thus cannot allow for transition. 16 does allow for transition; it is the gateway through which space and time appear. But this transition may not 'take form'. The transition can be expressed as a ratio, a precise relationship, a specified degree of change. We can imagine knowledge knowing how to operate these ratios and time conducting by means of the ratio (understood as the knowledge that time needs to move). Ratio connects 'from' and 'to', giving to the point the holding power it needs to manifest.

Zeros are replete with 16, but remain not-established. They can always be condensed back into the starting point. Even if this shift makes us lose perspective, nothing is lost. There is no difference between points, for space is space and time is time. All points are connected. The distinction between Old Zero and new zeros is just moves made by time and knowledge in space. Any point can be any 'from' or 'to'. Indeed, without this interpenetration, how could time and space unfold freely? No point is privileged.

Rationality likes a starting point. But the point is 16, and space and time are prior to logic, and ratios are prior to the perspectives that give 'here' and 'now' and 'there' and 'then'. Time unfolds, whether we go along for the ride or not. Any point can be opened up. "If knowledge knows how to operate the ratios, every transition is a new chance." (134)

Imagine 16 transitions for each of 16 transitions; imagine that 16s must interact to manifest. Precise connections unfold, because they must. There is movement, but not movement from here to there. Points manifest as reflections. 16s share 16s; time shares space. It can all unfold. "Faster than the speed of light is the motion of time." (136) The unit is a point, the point becomes space, and space is unified. Time moves in all directions.

Knowledge 1 141–55

Search for Knowledge K1

Knowledge is always changing, and not all change represents progress. We cannot assume that our current knowledge and ways of knowledge are the best available. Today it may be that time and space themselves have changed their qualities as well. Going to this fundamental level may be the best way to deal with the challenges facing us. We can challenge our 'according to' understanding in favor of a knowledge that is not 'for' or 'to' a subjective knower. But can we change this, when we are in the business of selling limitation? Yes, if we are willing to speculate and careful not to reach conclusions too early on, and if we aim for precision grounded in inspiration.

Ordinary knowledge depends on making the point, but the point is multiple: to be the one who knows; to know what is known. In making the point, we rely on space and time; proceeding from point to point, we assure that time and space will have certain structures. Points proliferate furiously, and knowledge circles round and round. If we could open each point as it arose, this momentum would not develop, and other forms of knowledge might become available. But this would require recognizing the point of knowledge as zero-point: as the prior of perception. This would mean opening the starting point: the knower, the 'I' of 'I am here'. This requires a zero-based kind of knowing. Perhaps we could open the angle of awareness, the cone that radiates from the point.

We know that knowledge accumulates in time. The knower knows, but what is the cone of time that leads to that moment of knowing? And what is the cone of knowing itself; or better, what are the multiple cones that radiate, depending on the orientation we take? More fundamentally, we can look at the structure of all such knowing: the baseline that serves as an obstacle to knowledge, constructed by space and time, and at the assumptions about space and time that operate in this model. If these characteristics changed, would space and time cease being obstacles? Would the cone be transformed? If we visualized the cone of knowledge in action in each moment, this could help. The prospect opens that knowledge could take on a time and space embodiment.

Knowledge 2 156–64

Structures of Knowledge K1

Once the baseline gives 'from' and 'to', appearance is determined by what can be measured, and limits on knowledge become central. Against a grid of space and time, solid objects take form. The ratio of knowledge and ignorance is fixed. But the projections we arrive at all take time and space for granted, and so may be inaccurate. Lower-level knowledge has great difficulty in noticing its own limits, but there are reasons for being suspicious.

The 'from-to' is basic to lower-level knowledge, shaping how we understand time and space and confirming the knower. But if knowledge had no baseline, no history of arising, new possibilities could open. The cone of knowledge is set up almost instantly, and as a result we never look at the distorting tendencies that shape the 'here'.

A point of origin cannot readily be a point of access for knowledge. Then can we question the point of origin: the 'from'? We can look at styles and structures of knowledge in different disciplines, which rely on 'from' in different ways.

Knowledge 3 165–74

Orders of Knowledge K1/K3

The self insists on its role, making the structures of the 'from' its own. But the order the self imposes has a habit of breaking down, of proving unstable. That order places all emphasis in the act of knowing on the subject, but in fact the object must be open to having an order imposed. Also, there is the time-space distance between subject and object, which can never be bridged. A steady cognitive friction leads to obscuration. The naming activity of language only confirms them. Language introduces the ratio between the old known and the new unknown, but also the ratio of the gap between being and the words we use to name it. There is a further gap as well, between the words we use and our knowing what they name (e.g., the zero-point). Still, it may be possible to arrive at a knowing that does not depend on the words used. At the same time, we can draw on the power of language. Dialogue with language has its own liberating momentum.

Imagine opening the cone of knowledge. It seems the point of origin will vanish. Does this mean losing perspective? Or can we engage the perspective of the whole, one that depends on no particular point? In that case, the glue that held the cone together—first level time and space—could come unstuck, making time and space and knowledge available throughout the whole. Now 'I am here' no longer confines us. Order appears, but it does not hold. Knowledge can become an expression of the time and space in operation. We can see time, space, and knowledge as sponsors of the ratio, giving an order that we do not impose, making available a new body of knowledge, the womb of transformation.

Knowledge 4 175–83

Cone of Knowledge K2

'Here' is taken as the zero-point for knowledge. It leads to distance from 'there', setting the limits for knowledge. The cone has its structures and limits that shape what can be known. But how are lines of directionality generated? They could be considered constructs formed by previous acts of knowing, creating the possibility of endless branchings.

The cone defines an angle of vision. But how can we return to the 360 degrees of the zero-point? We could recognize different cones, which still share the quality of belonging to larger cones, with shared factors. This suggests an interconnectedness of knowledge. Chemicals are connected, languages are connected, the senses are connected, etc. Then there is the connectedness of being connected; e.g., language and the senses in some sense point toward the same thing. The self as knower shares these connections; shares in the universal space-time-knowledge-cone. So there is the possibility of sharing in the cone of knowledge without being bound by the 'here/there' subjective cone. We could imagine this larger cone as made of light, or as returning to zero.

Time before the beginning is not bound by the cone. Is the same true for knowledge? It seems it could take any point as 'here'. Suppose we traded 'here' and 'there', or 'now' and 'then'? What then? For instance, subjectively we are free to reverse the temporal flow, and though this is only subjective, it has an impact on the solidity of the temporal dynamic. "If we return to the zero, perhaps time and space can go everywhere and all at once." (182)

Knowledge 5 184–94

Zero of Knowledge K2/S2

When zero becomes a point of view, the baseline takes form. Yet as the fullness of zero, 16 remains available. Zero is the holding power of space and the energy of time. As the 'ground zero' of knowledge, it makes possible locatedness in physical space. But suppose there is an inner space, like the prior of time: a space that accommodates physical space. Could zero be the transition to that second-level space? The axes of zero suggest as much: a womb through which appearance emerges. On that view, each point is a singularity, incorporating in its appearing the whole of space.

How is the more fundamental knowledge linked to second-level space turned into subjective knowledge? How does zero become 'something'? What is the 'from' of 'from'? There must be a point of transition. Zero must hold the whole: the right knowledge open to fullness. Appearance emerges as the ratio between this knowledge and ordinary patterns. It is here that the alternative structures of point and cone and baseline come into their own, offering a 'mathematically precise' transition beyond first-level structures. Existence does not have the same gravity; its 'is' enters the 'beneath' of appearance and the depth of space that zero holds open within each ratio.

Every point offers this zero-point structure along its own axes of access. Feelings are good examples: feelings of yearning, of owning, of being lost. Awareness offers this integration, but it is not the awareness of the observer (cf. the TSK Giant Body exercises); rather, it is the knowledgeability inherent in "just interactions and shining outlines."

Knowledge 6 195–203

Freedom of Sixteen K2

16 as fullness is available, in a way that zero, which is not there, is not. In conventional knowledge, sets such as 16 emerge only as the cone unfolds, but 16 is available at the outset in any zero-point. In 16, space stays at the center. It is the foundation for appearance and the allowing for transition. In each zero-point, through the transition of zero, the 'x' of the axis becomes 16, granting knowledge access. In conventional knowledge, this becomes the axiom that founds the order: the 'pointness' of the point. Time can unfold, while the point remains open. 16 is the secret source, offering the ratio and boundless transition. Appearance can appear without being claimed by an owner.

In ordinary understanding, we may not engage the whole of 16; may stop at 15, missing out even on the clarity that is the aim of first-level knowledge. It is like taking positions in a democratic regime without first having reached the decisive point. "Attuned to the fullness of 16, knowledge gains access to the ratio that resolves into right action." (200) Not to arrive at this fullness is to live out of focus. 16 is only an intermediate step in arriving at higher knowledge; nonetheless, we must reach the decisive point.

16 as decisive point becomes the knowledge-point—the point of transition, whose mechanism is the ratio. Here is a second-level openness that maintains the freedom of 16. The three-dimensional zero points toward the fullness of 16, but zero in the end goes further, pervading all appearance. We could say that the point completes itself in every moment.

Knowledge 7 204–16

Knowledge Taking Shape K2

16 unfolds in time and space, expressing the laws that govern how something arise from the point of origin. Ratios express such hidden structures, including a ratio of the whole (the logos). The ratio reveals other underlying orders, each unfolding difference in accord with its momentum. We could speculate that the ratio gives form to the axes, which transmit the gravity of the background, becoming the axioms of the order.

The ratio could become more accessible to knowledge, starting with the way the baseline shifts its shape in response to gravity. This is the ratio of experience, shaped by meaning and purpose. 'Taking place' means instituting a curve. The ratio operates in three ways, marking out appearance, and appearance depends on these operations.

At the second level, the ratio would open appearance to space and time; at the third level it could become available as the supporting 'orderness'. The first level finds in this a secret code, since it cannot reveal the ratio directly. To break through this limit requires looking at fundamental mechanics: point and axis, 16 and zero, etc. As the ratio becomes available directly, fixed positions do not hold. In the ratio of higher levels, we understand knowledge as intrinsic to the system; we engage time-space-knowledge as the threefold axes of three-dimensional knowledge. In time, the ratio of time to space to knowledge approaches unity.

We can study the ratios of past knowledge and future trends and events; the ratios of ordinary mind; the ratios of possession. This would have great value, but would still leave us short of 16: the ratio of the whole. The closer we come to the power to specify, the more we activate the power to transform. But this requires turning from the past (the order we impose) to discover the order implicit in appearance. Not letting go, we cannot make the transition, nor can we discover the interconnections implicit in 16. We stay on the surface and pursue the linear. What we need is an alternative geometry, a meaningful geometry. For instance, suppose that time were a sphere, and that the sphere were the openness of space, inseparable from the ratio of the cone.

Knowledge 8 217–37

Transition to Knowingness K1/K2/K3

Space and time and knowledge meet in the ratio of 1:1:1. But is this a ratio at all? In this open light of knowledge, do rhythm and gravity function? Or does the ratio merge into zero? The frame for lower-level knowledge is subjective, but when specific ratios fall away, subjective positions may not be required. We can move toward a knowledge of knowledge. "The turning point comes when knowledge expands to include the . . . 'I am here'." (219) Now the whole of 16 is available, for the only owner of knowledge is knowledge.

At times we arrive at a point of perfect balance. Each point is newly born; the ratio has not been formulated and the transition endures. In a sense we are beyond 16. We are at the point of choice: the wholeness of the circle, the zero point and point of origin. Transition is transformed, returning to 'before' the ratio that gives lower-level knowledge, arriving at the decisive point.

Before the axiom or the 'x', there is the zero, occupying without occupancy. Ratio gives all interaction, but there is also the origin of ratio. The transition that gives ratio leads back to the being of the point. Ratio marks a phase transition into this or that: the decisive point. Perhaps the source is a fourth dimension: "the giving place for the ratio of 16." (228)

Here there can be a moving in all directions simultaneously. Yet 16 does not move, even though the ratio is always in transition. We cannot enter this dimension of all possibilities—the qualities that would grant entrance are not our qualities. Yet we stand at the juncture of time, space, and knowledge. All is available.

Time, space, and knoledge are the 'here' of our being, and we can know the depth of this here, engaging and embodying. First-level knowledge, relying on the ratio for connections, will fall short. Pointing out misses the point; it is always 'off'. But perhaps we can engage without a ratio of subject to object. We can see the mechanisms in operation, and we can do this even in missing the point. "The echo is only an echo, but the echo is also sound." (234) Ratio can merge with no ratio.

There are three keys. All points are symbols of zero. We can experience the depth of being. Knowledge unites with the expanse of experience. In this third-level transformation, 'I am here' enters the depth of fullness. The embodiment of the Body of Knowledge becomes our embodiment. The secret dimension of the sacred is here.

Appendices

TSK Methodology

The TSK books contain many discussions of how to awaken inquiry and engage a new vision of reality. See, for example, many of the topics in the Knowledge division of the TSK Guide (Topics 51–75; also Topic 15 in the Space division); see also Unit One of the Study Guide. Other discussions also offer wonderful insights and suggestions. Here is a partial list, by volume.

Time, Space, and Knowledge

Love of Knowledge

Characteristics of the Books

This Appendix offers what can only be an impressionistic account of the six TSK books in terms of their qualities and flavor, their special strengths, their likely audience, and some obstacles readers may encounter in dealing with them.

Time, Space, and Knowledge: A New Vision of Reality

TSK is the original book in the TSK series. Published in 1977, it communicates a tremendous sense of excitement. The reader is invited to join in a journey of discovery that promises to lead into a whole different reality.

Several factors set TSK apart from later books. Perhaps the foremost is the way the book unfolds as a carefully guided journey with a definite starting point, path, and goal. See, for example, the opening of Ch. 3, p. 49, which refers to "the new view of reality which is gradually being activated." Another way to put this is that the book constitutes a sustained argument: a presentation that the reader is meant to follow along with, arriving ultimately at a particular destination; namely, a new vision of reality. Of the other books, DTS comes the closest to generating this same kind of momentum, and it is thus no surprise that TSK and DTS are the two best-selling books in the series. LOK is also a sustained argument, but serves more to clear away limiting views; at the end of the book, the reader has activated knowledge and is ready to start on the journey toward a new vision of reality (a journey that in third-level terms is no journey at all).

Related to the carefully designed progressive structure of TSK is the close integration in TSK of exercises and text, something not found to the same degree in any of the other books. The exercises are vital to the development of the path the book lays out, just as the text is vital to drawing out the implication of experiences with the exercises. Given this integration of text and exercises, it is somewhat surprising how many people tend to focus either on the exercises or on the text, but not both. Readers who contact Dharma Publishing or the Nyingma Institute often say that they greatly admire the text, but do not have much experience with the exercises, while others say that they do the exercises regularly, but cannot read the text.

The reason for this may have to do with issues of perceived competence. Some readers try the exercises but mistrust their own ability to go into them deeply, or else are frightened by their experience. Others never even try them, saying (for instance), "I've never been very good at visualization." Other readers may be fascinated by the experiences the exercises evoke but stop short when they come up against philosophically sophisticated arguments on the one hand, or seemingly paradoxical claims on the other. Fortunately, much of the text consists of comments to the exercises, and this helps bridge the gap that might otherwise open. Since readers who focus on the exercises usually read the commentaries, they are introduced to key ideas, while readers who focus on the text are able to imagine by way of the commentaries what it would be like to do the exercises, and thus do not have to give up on the experience of the vision. The integration of text and exercises thus works even when it does not work.

Despite the tight organization and dynamic thrust of TSK and the careful integration of the exercises into the text, the structure of the book incorporates a certain tension. The book is divided into three parts (Space, Time, and Knowledge), while the gradual unfolding of the vision towards its fulfillment proceeds in a unified manner, from the beginning of the book to the end. Thus, by the end of the Space section the reader has been introduced to the notion of Great Space at the level of theory, but is far from being able to "access" it experientially. For this to happen, the sequence of experiences that the exercises aim to evoke must be followed out through the final exercises. Only toward the end of the Knowledge section, then, will the reader who has been doing the exercises actually arrive at the level of experience where Great Time, Great Space, and Great Knowledge can manifest.

This tension is not a shortcoming. What it means in practice is that the serious TSK student will want to return to the exercises and readings from the early part of the book with the insights that come from doing the later exercises and readings. The text often suggests doing just that, and stresses the circular nature of its presentation (see TSK xxxii–xxxiii). Similarly, some of the discussions contained in the early part of the book (e.g., Ch. 4–5, and even parts of Ch. 1) proceed at a very advanced level, and may not become fully accessible until the rest of the book has been read and worked with experientially. A course based on the book, then, would not necessarily start at page 1 and work all the way through. In this context, it is worth noting the value of reading the other books for understanding TSK ; many people who could not get anywhere with TSK on a first reading found it clear and enjoyable after working fairly intensively with one of the other texts.

In any case, it would be wrong to view the process of activating a new vision or view of reality in strictly linear terms. For instance, any second-level experience, whether it focuses on space, time, or knowledge, brings the potential for third-level vision to manifest closer to actualization. Compare the discussion at the opening of Ch. 8, which points out that we cannot have second-level experiences of time unless we already have an acquaintance with Great Space. One interesting consequence of this circularity and interweaving is that the book will not let itself be inhibited by its nominal structure. For example, the most sustained discussion of Great Time comes in Part One, on space!

TSK is shaped by the truly radical possibility that we could invite a new vision of reality, a vision not owned by anyone, put forward by anyone, or realized by anyone. The other books share this commitment to the underlying vision, but there is a sense in which TSK is the most revolutionary, because it deliberately sets out to overthrow an established way of seeing. This is a part of what contributes to the sense of adventure that the book conveys: The world itself could be transformed. The same message is certainly implicit in the other books, but in TSK it stands in the foreground, beyond all possibility of denial. Compare this statement from KTS 111: "This does not mean that a new reality must be found, so that consciousness will have something to embody 'into'. Instead, consciousness can be transformed in embodying fully the knowledge that is already available."

Even today, TSK continues to be most popular of all the TSK books. Many people find it deeply exciting, nourishing, and fulfilling. Yet certain people are scared off by it, or find it uncongenial, and even many people who buy it end up not reading it. Among the groups for whom it seems not to be well-suited are the following:

☆ Readers not confident in their intellectual abilities. This is variable: Sometimes readers who consider themselves non-intellectuals are caught up in the exercises and commentaries, and are willing to skip over parts that they do not understand. But for some readers, the vocabulary, which often verges on or adopts the style of abstract philosophic inquiry, is simply an insuperable barrier.

☆ Readers intent on having interesting kinds of experiences, without a strong interest in inquiry. Such individuals can indeed have good experiences in TSK seminars, but they will not get very far with the book or the vision, which depends utterly on inquiry and is less concerned with stimulating transnormal experiences.

☆ Readers scared off by demands. The TSK book asks for stronger commitments than the other books. For instance, it asks the reader to do Ex. 1 "over a period of days or weeks"; again, it suggests doing Ex. 15 for five hours a day over a period of three weeks. And the instructions for Ex. 23 begin, "First do all the preceding exercises." It makes strong demands in another way, since the opening Giant Body exercises are visualizations that some people simply cannot do. In the early days of TSK, these demands were taken more seriously than they have been in recent years; nowadays, there is a sense that it is possible to "dip into" the book and also the exercises, doing a bit here and a bit there. This has the advantage of making the vision more accessible, but it has a drawback, in that the sense of a rigorous and demanding journey has faded. In this sense, TSK suffers the fate of any pioneering work.

☆ Readers skeptical about new visions of reality. The later books are more cautious in their claims than the first TSK book; as suggested above, it is possible to read them as saying that we can transform our experience, without claiming that reality itself will change (though to be fair, the contrast between 'reality' and 'experience of reality' holds only at the first level). TSK is very blunt on this point, and it will invite blunt responses from skeptics. Some people may also be put off by what they may see as a certain arrogance. For instance, the discussion in Ch. 4 could be read as saying that religions can serve a useful function as adjuncts to TSK. Similarly, the claim that Great Space can reconcile science and religion is at once exciting and (for a skeptic) absurd; it invites the response that such reconciliation holds only at a naive level. Ch. 6 also makes the provocative claim that the closer science gets to the TSK vision, the more accurate it will become. Such statements are red flags for a certain kind of hostile critic, even if they have nothing to do with the validity of the TSK approach. In any event, the later books are far more moderate on these points.

☆ Intellectuals who believe that truth or knowledge can come only through rational arguments and/or the evidence of the senses. TSK certainly offers a good many rational arguments and distinctions and calls for close attention to sense experience as well, but it ties these in closely to other kinds of experience that rationalists or empiricists would simply not accept as giving rise to valid truth claims. And it does not aim at philosophical refutation, pointing out that the methodology it relies is one that an analytic philosophy will simply reject. For many individuals with a philosophical background TSK is thus likely to be irritating. On the other hand, if such a person is open to other ways of thinking, TSK may prove fascinating, because there is a sense that key philosophical issues are being addressed (see especially Ch. 4). In a sense, what TSK offers is an unusually satisifying version of philosophy, one that integrates intellectual analysis with experience in a way that leads toward transformation of human being. Compare TSK's call for "an integrated, natural intelligence, unfragmented into reason, emotions, sensations, and intuition." (TSK xxxiv)

☆ Readers unwilling to suspend disbelief. The TSK book, in contrast to some of the later books, starts

> right in with notions that are difficult to accept on the ordinary level. A prime example is found at TSK 10: "Space is projecting space into space!" On careful reading, this is not such a radical claim; in fact, it could be regarded as a level-one statement. But it throws some people off, and sometimes people put the book down at that point, as being "just too strange."

Most people who are drawn to the TSK vision by reading the first book remain connected to the vision through that book, and many long-term students see little reason to work with the later books. However, given that the other books in the series have also been available for many years, readers are growing more familiar with them, and in many cases are introduced to the vision through those books. For them, it is these books that embody the "true" TSK.

For some, TSK has been a wonderful starting point for the vision. But, as noted above, it makes strong demands on its readers. When TSK book was published, nearly thirty years ago, there was more reason to believe that readers would be willing to meet those demands. It was a time when it really did seem that reality itself might change. Today that sense of revolutionary potential has dimmed, and this makes the first TSK book perhaps less accessible than it once was. On the other hand, readiness for transformation seems to move in cycles, and it may be that we are entering a cycle in which the strong voice of the first TSK book can be heard once more. If so, readers may once again be prepared to approach the TSK vision, as presented in TSK, as a unified whole, a journey to be followed from beginning to end.

Love of Knowledge

Like TSK, LOK unfolds as a carefully structured journey. The LOK journey is more linear than in TSK, more in accord with lower-level knowledge, and for these reasons some readers find it a less bold and challenging work. Another way to say this is that LOK focuses strongly on a critique of lower-level knowledge, and is relatively oblique in its presentation of Great Time, Great Space, and Great Knowledge (three terms that are never used in LOK). For those well-schooled in Western forms of knowledge (for instance, through science or the more theoretical disciplines), and especially those trained in academic forms of inquiry, LOK can be the best introduction to the TSK vision. TSK presenters working with this kind of audience often turn to LOK, because it can be read in sustained sections without 'erupting' into passages or terminology that would lose a mainstream reader.

But this is only one way to tell the story. It is just as true to say that LOK begins with a leap directly to the third level of knowledge (as SDTS begins with a leap to the third level of space and time), and that its first-level style of presentation is simply a way to open up the possibilities for time and space. Seen in this way, as an exposition of knowledge that goes beyond the presupposed, it is truly a tour de force.

LOK proceeds in a logical, rational manner, making no unusual claims but gradually deconstructing many of the assumptions we take for granted. Readers will find parallels to themes in Western thought that can provide helpful entry points, and a language consistent with academic discourse. This is not to say that LOK is a "conventional" work, but it makes enough concessions to conventional approaches to keep a more cautious reader engaged.

Where TSK starts with space and time (as do KTS, DTS, and SDTS), LOK starts with knowledge, and knowledge remains its primary focus throughout. More basic than this difference, however, is the shared quality of appreciation that shines through both books from the outset: appreciation for space in TSK and for knowledge in LOK. To see this, compare the first sentence of TSK and LOK:

TSK Whatever we do, wherever we go, whatever happens on this crowded surface of interactions constituting our world, there is also the sky.

LOK No matter who we are or how we live, our lives are shaped by what we know.

In Part Two of LOK, time becomes a central theme. Space, though certainly not neglected, never receives the same degree of attention. Some readers report being profoundly moved by the commitment to openness and inquiry that suffuses LOK; on the other hand, it seems fair to say that many readers find it easier to connect to time and space directly than to let knowledge activate time and space by showing (see Ch. 33 and 391–92) how analysis can embrace and embody.

LOK emphasizes the role of the self more than any of the other books in the series (see especially Ch. 5 and Part II), and the self is a topic sure to interest anyone. However, its approach to the self is more philosophical than psychological, and many people lack the training and background to enter into this kind of discussion easily. For this reason, it has proved fruitful in recent years to combine study of LOK with readings from the closing chapters of *Knowledge of Freedom*, a book that does not form part of the TSK series but deals with similar themes. *Knowledge of Freedom* takes a more psychological approach to the issues of the self, and complements LOK nicely. It could be read as a prequel to LOK, and in fact this accurately reflects the history of the two books' composition (*Knowledge of Freedom* 1984; LOK 1987).

The exercises in LOK are not integrated into the text in the same way as in TSK. Instead, they are presented as a possible resource for inquiry (like the drawings and captions that begin each chapter). Organized into series that are distributed throughout the book in a non-linear way, the exercises are an invitation to the reader to carve out a personal path of inquiry. Some readers prefer the more structured approach of the TSK book, and also appreciate the commentaries to the exercises in TSK, which do not appear in LOK.

This is not to say that the exercises in LOK are unimportant or irrelevant to the text. It would be a useful exercise to try writing commentaries for the LOK exercises that showed how they advance the argument of the book, on analogy to the way the exercises are integrated into the text in the TSK book. In any event, the LOK way of presenting the exercises makes a demand on the reader to be deeply familiar with the structure of the book and the way the exercises relate to that structure. It invites the reader to return to LOK again and again, carving out a new path each time, learning to use LOK and become expert in it as one would become expert with a tool.

This shift in the way the exercises are presented parallels another, more fundamental shift from TSK to LOK: a shift in the nature of inquiry. In TSK, inquiry aims at a result: 'putting in place' a new vision of reality. In LOK, on the other hand, inquiry does not aim at a specific result; rather, inquiry itself becomes an exercise: a way of evoking the new vision. LOK inquiry is not 'directed' in the same way; it is "free and open." The difference can be expressed by comparing TSK's "When Time and Space are directly appreciated, that is automatically knowledge," (216) with LOK's "Knowledge is known in the act of knowing." (300)

On the other hand, it is possible to overstate this difference. TSK suggests (xxxiv) that "if . . . an open intelligence is brought into play in reading this book, even the reading and thinking process itself can become a visionary path." LOK (especially Part III) could be read as expanding on this possibility, which also receives extended treatment in VOK and DTS.

Knowledge of Time and Space

In the preface to KTS, Tarthang Tulku remarks that the book may be more accessible to readers than TSK, because the language is "less technical" and "more accessible to the general reader." In a sense this is true: the language of KTS is used in a variety of ways: at times playful, at times poetic; at times rigorous and rational; at times evocative. But the overall effect is to make the book more challenging, not less. KTS frustrates readers' expectations: It does not proceed in an orderly, structured way, or present one coherent argument; it does not make the same extensive use of exercises that is found in TSK and (in a different way) LOK. It takes up and then drops one theme after another; it is loosely woven rather than tightly knit (one indication: the chapters are not numbered, except in the table of contents). Rather than a steadily burning flame, it is a beautiful explosion of a book, like a fireworks display. Perhaps more than any of the other books, it is inspired by the third-level of the TSK vision: not just in the language it uses, but in its outlook and approach. For this reason, it manifests a certain 'strangeness' that makes it difficult for some people to read. On the other hand, the beauty of the vision as it finds expression here can touch people who would not be moved by analysis or be likely to have 'profound' experiences in doing the exercises. It is perhaps worth noting that the title for this book comes from a passage in KTS.

KTS is more idiosyncratic in its use of language than any of the other TSK books (matched most closely in this regard by DTS and SDTS). Countless terms are used in unfamiliar ways; countless ideas are introduced that stimulate either wonder or confusion, depending on one's frame of mind, but that are certainly not treated "rationally" in any usual sense. Still, most of the book—perhaps 90% or more—can be read as making sense on the conventional level. And many of the themes that KTS introduces (for instance, the field, or the zero-point) add wonderfully precise dimensions to the exploration of the vision that are taken up again in later books. KTS also features an extremely down-to-earth treatment of sense experience that helps reconnect this aspect of human embodiment to the vision. This ties in to the appeal that the book seems to have for artists. Compared to the rigorous argument and presentation of LOK, KTS explores a completely different aspect of the TSK vision: an alternate universe of TSK inquiry.

KTS is by far the longest of the TSK books, but its chapters are by far the shortest. This juxtaposition makes it natural to read the book in bursts, rather than from beginning to end. This tendency is further encouraged by the organization of individual chapters into sections that explore facets of a particular theme, like looking at a polished gem from different directions. Within the major parts of the book, there is a loose progression from lower-level time, space, and knowledge toward Great Time, Great Space, and Great Knowledge, terms which here re-emerge after dropping away in LOK. The fourth section, on Intimacy, takes the practice of appreciation, emphasized in both TSK and LOK, to a new level.

Except in the section on space, KTS does not have exercises. But here the promise of the first TSK book that the reading of the TSK can itself become an exercise is realized in a wholly unexpected way. Chapter by chapter, KTS presents the beauty of the TSK vision, challenging the reader to respond. It presents the mechanisms of existence, the "esoteric biology" of the universe. (91)

In KTS, 'levels' take on a different significance. In TSK, making the move to level two meant engaging experience more directly. But in KTS, the text itself is a level-two telling: a different knowing struggling to find expression. A powerful impulse bubbles forth in words, and the chapters are sometimes incredibly dense, as though they contain more than words can express. Repeatedly the text retells at the second-level the first-level logos, a reformulation stunning in its alterity. It asks if we are "willing to abandon the structures of conventional understanding?" (91) For more on levels in KTS, see Ch. 95 and 97, and consider p. 438: "Second level analysis can be understood as a concession to first-level understanding."

All the above holds most strongly for the discussion of time. The discussion of space, at least to begin with, is far more measured. It inquires in a more accommodating way, gently opening the level-one perspective on space.

Second-level space is treated in a new way in KTS. In one aspect, at least, second-level space is the 'field', the space that allows physical space (see Ch. 31) Compared to TSK, this constitutes a shift. (Compare, however, the suggestion in TSK that physical space is the "local representative" of space.) Then again, second-level space is introduced in KTS as something "prior to all interpretive structures," which seems to set it apart from the field. This suggests that second-level space 'gives' the field in much the way that second-level time gives the first-level temporal order; it also suggests that second-level spaces can multiply endlessly. Perhaps it is right to say that the field is neither first-level nor second-level, but in between. See in this connection the highly illuminating discussion of the relation of knowledge to space in Ch. 35.

To start exploring the TSK vision with KTS would be a tremendous challenge. But approached in the right way KTS is a remarkable resource for anyone, no matter what his or her level of involvement with TSK. It reveals how utterly transforming the vision might be. And it offers countless openings, for every chapter (or series of chapters) is a way into the whole of the vision.

Every one of the books has its own way of implementing, rather than simply expressing, the TSK vision. In KTS it happens on each page: almost, sometimes, in each sentence. It is not something that builds slowly, or that is best engaged as a whole. Instead, it shows up like a strange visitor from another planet. You almost expect the pages to move beneath your gaze.

Visions of Knowledge: Liberation of the Modern Mind

VOK is the shortest of the TSK books: leaving aside the sixty questions with which the book begins and the back matter, it is only 133 pages. It is also the only book by Tarthang Tulku that appears in the series called "Perspectives on TSK," rather than the TSK series itself; as he writes (ix), it constitutes his response to the expression of interest in the vision that the other books of the Perspectives series manifest.

Partly for this reason, VOK has perhaps not received the attention it deserves. The book focuses very strongly on the nature of the knowledge we rely on, its links to time and space, and on the possibility for activating new kinds of knowledge. It introduces imagination and speculation as valid and powerful tools for TSK inquiry, a theme continued in DTS and especially in SDTS. Read on its own terms, VOK can make a wonderful introduction to the TSK vision, and in particular to the practice of inquiry as life-transforming.

Unlike the books that appear in the TSK series itself, VOK does not present a complete vision of TSK. But this does not mean that it lacks coherence or unifying themes. The theme of VOK is inquiry, and in this respect it links up with and extends the inquiry of LOK—as always, in unexpected ways. It also has its share of new themes and approaches; for instance, the introduction of the terms "trans" and "trance." These ideas are not explored in detail; instead, VOK invokes the sense that imagination and speculation can lead directly to knowledge without the need for a linear path.

The sixty questions at the start of VOK have not yet attracted systematic inquiry. It seems the right way to work with them would be to integrate them into your own thoughts over a relatively lengthy period of time. This does not necessarily mean making these questions into your own speculations, though that would be one way to proceed. The alternative would be to let these questions become the gateway to others; to stimulate what LOK calls a "rain of questions." Then the link between questions and inquiry might be activated in a more comprehensive way.

VOK insists that inquiry must be precise (71); that we need a true methodology. There are many ways to break through conventional knowledge—the religious and mystical, art, and so forth—but without precision each of them risks getting caught in repeating cycles of time, even when time itself is in some sense transformed. Such precision is always possible: no approach is excluded. (116)

VOK has its own version of the second and third level. The viewpoint it takes on knowledge is inspired by the third level, by Great Knowledge, but the precise path of inquiry it unfolds could be considered its second level. In its particulars, this path looks closely at time and space in their relation to knowledge: an ongoing inquiry on the way to Great Knowledge. To follow this path, we have to recapture the significance of time and space and knowledge; to see how our present world arises through a TSK interplay. (89) We realize this not abstractly, but in experience. In VOK, questioning itself serves as the vehicle for evoking such experience.

Dynamics of Time and Space: Transcending Limits on Knowledge

Many people have found DTS a good place to start their exploration of the TSK vision. It begins with a gentle meditation on the nature of space and its relation to substance, based on a natural and inviting thought experiment. From there, it gradually but also rigorously unfolds the space dimension of experience, challenging the limitations we usually impose unthinkingly. Soon readers are reflecting on topics that go far beyond the presuppositions that ground conventional experience.

This combination of rigor on the one hand and gentle invitation on the other is characteristic of DTS, in contrast to the hard-charging dynamic of TSK. DTS is also gentle at a deeper level. Whereas TSK challenges the nature of the reality we experience (for instance, discussing at some length whether it is possible to walk through walls), DTS directs its challenge at our *experience* of that reality. Of course, the conventional distinction between reality 'as such' and reality as we experience is called into question by the TSK vision at the deepest level, but the difference in emphasis seems real.

Like VOK, DTS has its own approach (and own vocabulary) for the second level of the vision. Among the terms it introduces at this level are field communiqué, eknosis, 'happeneded', etc. Like KTS, DTS seems to adopt the strategy of using conventional terms (or at least terms derived from conventional terms) as interim structures for deconstructing lower-level knowledge.

Also like KTS (and despite the rigor of its analysis in places), DTS is very free in making imaginative leaps that upend whatever we thought we already knew about time and space. A good example is the sudden suggestion (42–44) that one fruitful approach to the 'zeroless' is to 'flatten' space.

Another interesting characteristic of DTS is its sense that the third level of space, time, and knowledge is to be found 'underneath' or 'outside' what can be put into words. See for example, the whole of Ch. 14: Opening Unknown Time. There and elsewhere, DTS draws freely on the power of speculation, far more ready to question its own presentation than to answer the questions it raises (See, for example, p. 136: "Why call this unknown structuring element 'time' at all?") Quite often in such contexts DTS presents level two as a way of enacting level three, of making it available within the ordinary. For some readers, this readiness to open even the presentation itself to inquiry may be frustrating. Like TSK, DTS operates on the basis of a unified vision that gives rise to a unified vocabulary, yet the vocabulary itself is always changing, always opening in new directions. This accords with the emphasis on improvisation in Ch. 15.

In a sense, Ch. 14–16, which close out the first half of DTS, could be seen as a turning point in the DTS presentation, marking the transition to a way of exploring the vision that marries the unified vision of TSK with the sense of inexpressible play so richly explored in KTS. In Part Three of the book, which follows immediately, a wealth of different images suceed one another, playing off each other and suggesting ways of being that words cannot describe. A good example is the discussion of light that begins in Ch. 19.

Unlike TSK, where the exercises are integrated into the text, the exercises in DTS have been placed in their own section (Part Four: Transformations). Yet the exercises unfold in much the same way as in TSK: first a context, then the exercise, then a commentary that deepens the exercise, taking the presentation in new directions. It is as if Part Four of DTS is a second 'telling' of the first three parts of the book, based this time in experiential inquiry.

The net result is to encourage an approach that moves from the exercises in Part Four back into the presentation in Parts One through Three and then back again in an expanding spiral. All this amounts to a working out of the approach described in the opening pages of LOK (xxix):

> The style of inquiry introduced here counteracts the tendency toward 'singling out', thus signaling a return to balance. With no positions or possessions, such inquiry is committed only to knowledge, and this commitment, *because it remains available as a subject for inquiry*, is in an important sense self-correcting.

But we do not have to turn to LOK for a description of the special flavor of DTS. We can find such a description in DTS itself. For it seems that this book works out in a remarkable way the invitation contained in its preface (xxiii):

> I do not own this vision. You are free to make your own commentaries, add your own insights, develop your own versions . . . I say this as more than a polite invitation. It is really up to you to activate this vision in your own life.

Sacred Dimensions of Time and Space

Readers of the books in the TSK series grow used to the idea that each new book will introduce new ideas and approaches and surprising twists on the TSK vision. Even in light of such expectations, however, *Sacred Dimensions* marks a break with the previous books in the TSK series, a completely different approach to the TSK vision. While there are parts of SDTS that accord beautifully with the earlier volumes (the Orientation, for instance [xix-xxxix] is one of the finest introductions to the TSK vision to be found anywhere), many aspects of the book enter into uncharted waters, introducing concepts (e.g., '16'; the ratio) that have no parallel elsewhere.

What is going on in this radically new presentation, and how can we respond? For an answer, we can turn to a suggestion made in the text:

> It is almost as if space has been given a voice. Attuned to its presence, we make ourselves at home, waiting to hear what it has to say. (25)

If space itself 'speaks' in SDTS, it is not so surprising that its message sounds unfamiliar to our ears. We are used to a style of presentation in which the self and the activities of the mind play the central role. The centrality of the self is challenged throughout the books of the TSK vision, but now the centrality of mind is challenged as well. Again, the text is clear on this point:

> In this description, we do not have to assign a direct role to mind at all. Instead, it may serve us best as a compelling analogy. The mind is open for anything, ready to receive all content. Is the openness of space, as symbolized in the 16 and also in the zero-point, akin to this openness of mind? Is it different? (31)

It is as if the founding story of the self, explored in LOK—the conductor's story of DTS—were abandoned, leaving us with none of the presupposed structures we ordinarily use to make sense of experience.

SDTS is rooted in a level-two reversal of our usual approach to space, time, and knowledge. It retells the content of experience in a way that subverts what is told. Offering form without substance, it can seem at times to be conveying cryptic messages from a parallel universe, in which the usual human concerns have never taken form. Put differently, SDTS offers the zero of space as form rather than space. A possible analogy is the 'just interactions' approach of TSK: in a sense SDTS works out the geometry and geography of that view.

The SDTS perspective is certainly challenging: a suggestion that zero can build up appearance and reality without ever having to rely on substance. To help communicate this wholly non-standard view, the text relies at times on images of a surprising poetic beauty. Here, for instance, is a passage from the end of the space section:

> Beneath the surface of separate appearances, everything in the universe connects to space, heart to heart. All edges are united by deep lines of space. Imagine the still of night, with everyone asleep; suddenly, through magic, the roof of every house is lifted off—all imagined spaces merge, all the secret thoughts flow into one.

An Alternate Study Guide

Here is a study guide I prepared more than a decade ago, before publication of SDTS. Some readers may enjoy tracing out the connections its suggests.

I. Starting Point: Our Present Understanding as Limit and Challenge

A. The Present Situation for Knowledge

1. Breaking through restrictive views: TSK xxviii–xlvi
2. Beyond fruitless ways of being: LOK xxxix–l; LOK Ch. 1, 8–12
3. Integrating being: KTS xiii–xxi
4. Technological knowledge: LOK Ch. 3–5; cf. Ch. 20; VOK 78–83
5. Openness of inquiry: VOK vii–xix, 3–13, 55–66, 69–72
6. Activating knowledge: DTS xxv–xliv

B. The Prospects for Fullness: STK Interplay

1. TSK Ex. 7, 9A
2. LOK Ch. 12, 26, 48; Ex. 1, 10–12, 34, 47–50 (lightly)

3. KTS Ch. 44–46, 64, 68, 86, 110
4. VOK Ques. 38–40, VOK 72–76, 131–35, 152–55, 165–77
5. DTS 182–83

C. Opening Space

1. TSK Ch. 2, esp. 31–46; Ch. 4, esp. 91–98, 111–14; Ex. 7–10, 15
2. LOK Ex. 27
3. KTS Ch. 25–32, 35, 49–50
4. VOK Ques. 1–15
5. DTS Ch. 1–2, 4–6; 65–68; Ex. 7A, 8–9

D. Presence of Time

1. TSK Ch. 6–7; Ex. 23–24
2. Linear time: LOK Ch. 15, 21–23; Ex. 3, 13–14
3. Unknown time: KTS Ch. 1–3
4. Other times: VOK Ques. 26–30
5. DTS Ch. 9–11; DTS 125–44, 285–89; Ex. 7B, 12

E. Invoking Knowledge

1. TSK Ch. 10–12
2. LOK Ch. 2, 48; Ex. 46
3. KTS Ch. 57, 63–65, 85
4. VOK Ques. 20–25, 37, 41–42; VOK 77–78, 91–93, 98–102
5. DTS Ch. 17–18, 24; DTS 277–78

II. Path

A. Presence

1. Body
   a. Embodiment: TSK Ch. 2; Ex. 1–6, 23, 34
   b. LOK Ex. 3, 24
   c. KTS Ch. 40, 98

2. Senses, feelings, emotions
   a. TSK Ex. 7, 24, 35
   b. LOK Ex. 4, 6, 17, 19–20, 28–31, 34–39, 41, 44
   c. KTS Ch. 47–48, 83–84, 87–89
   d. DTS 210–17, 245–47; Ex. 4, 6, 9B–C, 20, 22
3. Thinking and the nature of mind
   a. TSK Ch. 2, 11; Ex. 11–14
   b. LOK Ex. 2, 4, 6–9, 39–40, 43, 45
   c. KTS Ch. 7, 33–36, 44, 72, 101, 107
   d. VOK Ques. 43–46; VOK 123–31, 138–43, 147–52
   e. DTS Ch. 3; DTS 49–70, 245–47, 258–60; Ex. 1–6, 9C, 10, 16B

B. Challenging Limiting Structures

1. Polar knowledge
   a. LOK Ch. 13–14; Ex. 16
2. Bystander/positions, models
   a. TSK Ch. 3–4, TSK 98–110; Ex. 16, 28
   b. LOK Ch. 6–7, 10–11, 13, 16–17, 30–33, 38; Ex. 16, 38
   c. KTS Ch. 4–5, 8, 37, 67–69, 73–75, 90–91
   d. DTS Ch. 17–18
3. Structures of the Self; Time and Experience; the Witness
   a. TSK Ch. 8; Ex. 17, 25, 30
   b. LOK Ch. 5, 13, 17–27, 44; Ex. 1, 3, 5
   c. KTS Ch. 4–6, 8, 66–67, 75
   d. VOK 138–143
   e. DTS 49–50, 296–98; Ch. 9–10, 12; Ex. 5, 11–13, 17, 23C
4. Language and narrative; naming; truth, linearity
   a. TSK Ch. 14; Ex. 18–23, 27
   b. LOK Ch. 17–18, 21, 25–27; Ex. 13–15, 18, 21–23, 25–26, 32, 33

   c. KTS Ch. 4, 7, 9, 10, 62, 78–81, 105
   d. VOK Ques. 47–60, VOK 58–63, 93–98, 105–12, 120–23
   e. DTS Ch. 3, 8, 10, 13, 18; DTS 160–64, 258–60, 285–89, 310–12; Ex. 2, 4–6, 11–13

5. Logos, temporal order, communiqué
   a. LOK Ch. 27–29
   b. KTS Ch. 8, 11, 14, 18, 24, 57, 92–94
   c. DTS Ch.2–3, 13, 18

C. Inviting Transcendence

1. Inquiry and magination
   a. LOK Ch. 33–36, 47
   b. KTS Ch. 61–63, 73, 89, 111–112
   c. VOK Ques. 31–36; VOK 57–66, 67–69, 104–5, 119–20, 126–28, 138–43, 161–63
   d. DTS 65–70, 189–98, 203–5; Ex. 17–18
2. Read-Outs and Pointings
   a. TSK Ch. 9; Ex. 26
   b. LOK Ch. 33, 38, 45, 46; Ex. 14, 22–23
   c. KTS Ch. 11–12, 18, 21–23, 53, 90, 105–9
   d. VOK Ques. 52–55; VOK 112–117
   e. DTS Ex. 13
3. Momentum; dynamics, rhythm, tracing, conducting
   a. KTS Ch. 8–9, 12–21
   b. DTS Ch. 11, 14–16, 18; DTS 321–24; Ex. 13–16, 19
4. Multiple dimensionality; field
   a. LOK Ch. 49
   b. KTS Ch. 31, 37–46, 51–54, 70–72, 82–84
   c. DTS Ch. 8, 14, 18; 315–19
5. Knowledge, identity, opposition, dichotomy
   a. LOK Ch. 5, 39, 44, 47; Ex. 5, 39, 41–42

   b. KTS Ch. 10, 18, 53–56, 72
   c. VOK Ques. 16–19
   d. DTS Ch. 3, 5, 10, 12; DTS 142–144; Ch. 15, 21; 267–271

6. Zero, nothing, negation, not-knowing, 'x'
   a. LOK Ch. 40–43, 45, 47
   b. KTS Ch. 1, 19–21, 26, 35–36, 53, 58–59, 73–78, 104, 109–10
   c. VOK 157–163
   d. DTS Ch. 1–3, 6–8, 15

7. Light
   a. TSK Ex. 32–34
   b. KTS Ch. 86
   c. VOK 136–138
   d. DTS 166–69; Ch. 19–21; Ex. 5, 9–10, 21–22, 25

III. Goal

A. Presence, Intimacy, Magic

1. TSK Ch. 7, 15; Ex. 25, 33
2. LOK Ch. 37–38, 48; Ex. 33, 37
3. KTS Ch. 45, 55, 80, 88
4. VOK 112–117
5. DTS Ch. 18; Ex. 1, 10, 12, 16B, 17, 20–21

B. Not-Knowing, Great Knowledge, Body of Knowledge, Being

1. TSK Ch. 11–16; Ex. 27–29, 30–35
2. LOK Ch. 11, 45–50; Ex. 45–50
3. KTS Ch. 20, 60–61, 64, 71, 91–112
4. VOK 83–90, 143–45
5. DTS 91–94; Ch. 21–24; Ex. 23–25

Articles by Tarthang Tulku on the TSK Vision

This brief list, presented in chronological order, excludes articles in the Perspectives on TSK series (specifically, in Dimensions of Thought *and* Light of Knowledge*). It also does not list excerpts from the TSK books that have been published elsewhere.*

"A New Vision of Reality." In *Buddhist Thought and Asian Civilization: Essays in Honor of Herbert Guenther on his Sixtieth Birthday*, 268–279. Ed. L. Kawamura and K. Scott, 1977.

"Time, Space, and Knowledge: Interview" *Gesar* 4(3): 5–11, 1977.

"Time, Space, and Knowledge: Interview." *Gesar* 7(4): 31, 1982.

"Beliefs: A TSK Perspective." *Gesar* 8(1): 35–37, 1984.

"Time to Open Knowledge." *Gesar* 10(2): 12–13, 1990.

"Partners with Knowledge." *Gesar* 10(3): 2–3. 1991. Reprinted in *Teachings from the Heart*, 156–58, 1998.

"Time, Space, Knowledge: An Interview with Tarthang Tulku." In *Teachings from the Heart*, 209–221, 1998. Originally published in *Gnosis: A Journal of the Western Inner Traditions* 39, 44–48, Spring 1996.

"I and Zero." *Gesar* 14(2): 18–20, 2000.

"Opening a New Perspective: A TSK Creativity Exercise." *Annals of the Nyingma Lineage in the West 6(2): Creation of the Copper Mountain Mandala of Odiyan*, 546–49, 2002.

"Restructuring the Bankruptcy of Knowledge." *Annals of the Nyingma Lineage in the West 6(2): Creation of the Copper Mountain Mandala of Odiyan*, 558–62, 2002 .Perspectives on TSK Series

Books of the Perspectives on TSK Series

Dimensions of Thought: Current Explorations in Time, Space, and Knowledge

Published in 1980, this two-volume set of thirteen essays by early readers of Time, Space, and Knowledge includes contributions from a number of well-known philosophers, phenomenologists, and psychologists, as well an extensive interview with Tarthang Tulku that offers one of the best short introductions to the TSK vision. Not surprisingly, these early essays tend to emphasize comparisons with other thinkers and approaches. Volume 2 also contains numerous short reports by readers. Each volume is available separately, Vol. 1, 237 pp.; Vol. 2 , 293 pp. ISBN 0-913546-79-8 (vol. 1); 0-913546-80-1 (Vol. 2).

Mastery of Mind

This set of eight articles by long-term students of TSK shows a shift away from theory toward applications. These are essays in the literal sense of the word: attempts to come to terms with fundamental issues. Their integrity and their respect for the TSK vision are reminders of how much the vision has to contribute to knowledge in all its aspects. 380 pp., paperback ISBN 089800-245-1.

Visions of Knowledge:
Liberation of the Modern Mind

The principal parts of this book were written by Tarthang Tulku, in acknowledgement of the work that went into Mastery of Mind. Exploratory, dynamic, and richly suggestive, Visions of Knowledge draws on the power of imagination and speculation to transform standard ways of knowing. Light and accessible, it invokes a transforming intelligence that delights in the riches available within ordinary experience. Part One offers sixty questions, with related commentary, as tools for inquiry. Part Two consists of short, related essays. Also included are comments by students and an analytic outline of Knowledge of Time and Space. 177 pp., 4 color plates, appendices. ISBN 08900-258-3.

Light of Knowledge: Essays on the
Interplay of Knowledge, Time and Space

Eleven articles, including an illuminating essay by Tarthang Tulku on "Geographies of Knowledge." Mostly written in an academic style, these articles suggest deepening appreciation for the TSK vision and the role it could play in our culture. 444 pp., ISBN 0-89800-287-7.

A New Way of Being

Twenty-three practitioners offer an inviting introduction to the TSK vision by showing how it unfolds in their own practice and thinking. Some apply TSK to their professions or fields, others engage in dialogs transcribed from study groups. Each article conveys a commitment to go beyond the limits of conventional knowledge, language, and concepts. 292 pp., ISBN 0–89800–356-3.

A New Kind of Knowledge

An engaging collection of seventeen wide-ranging articles by students who approach the underlying relation of time and space to knowledge from different perspectives, using a variety of styles. The tone is both friendly and invigorating. The articles contain penetrating reflections on knowledge, time, and experience, as well as exploring practical applications in psychology, teaching, business, and time management. 300 pp., ISBN 0–89800–357–1.

ISBN 913546-08-6

ISBN 089800-206-0

ISBN 089800-138-2

ISBN 089800-266-4

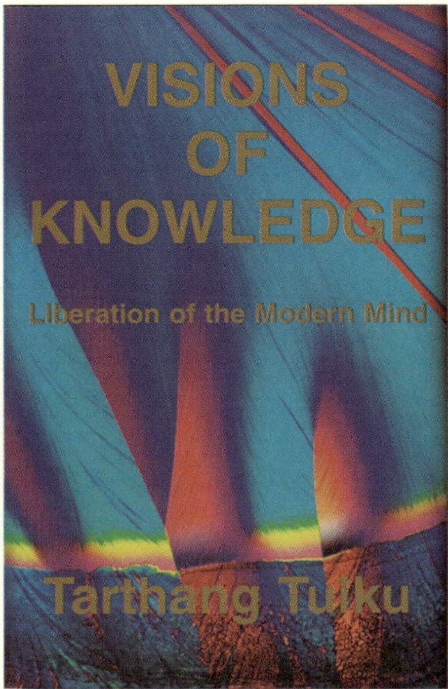

ISBN 089800-258-3

ISBN 089800-360-1

ISBN 089800-287-7

ISBN 089800-245-1

ISBN 089800-357-1

ISBN 089800-356-3